STUDENT STUDY GUIDE

7E

LINEAR ALGEBRA

With Applications

STEVEN J. LEON

University of Massachusetts

PEARSON

Prentice
Hall

Upper Saddle River, NJ 07458

Editor-in-Chief: Sally Yagan
Supplements Editor: Jennifer Urban
Executive Managing Editor: Kathleen Schiaparelli
Assistant Managing Editor: Becca Richter
Production Editor: Allyson Kloss
Supplement Cover Manager: Paul Gourhan
Supplement Cover Designer: Joanne Alexandris
Manufacturing Buyer: Ilene Kahn

© 2006 Pearson Education, Inc.
Pearson Prentice Hall
Pearson Education, Inc.
Upper Saddle River, NJ 07458

Printed in the United States of America

10 9 8 7 6 5 4 3 2 1

ISBN 0-13-193623-9

Pearson Education Ltd., *London*
Pearson Education Australia Pty. Ltd., *Sydney*
Pearson Education Singapore, Pte. Ltd.
Pearson Education North Asia Ltd., *Hong Kong*
Pearson Education Canada, Inc., *Toronto*
Pearson Educación de Mexico, S.A. de C.V.
Pearson Education—Japan, *Tokyo*
Pearson Education Malaysia, Pte. Ltd.

Contents

Introduction

This study guide is designed to accompany the seventh edition of *Linear Algebra with Applications* by Steven J. Leon. Each section of the study guide corresponds to a section of the book.

The purpose of this guide is to help readers learn and understand linear algebra. Generally linear algebra is the first course in the mathematics curriculum where the primary emphasis is on understanding mathematical theory. As a result you should expect to find the course challenging. However, if you work hard and see your instructor for help whenever necessary, you should be able master the course material. My aim in this study guide is to provide readers with the same sort of advice and help that I give to students during my office hours.

Generally, when students come in for help, I sometimes solve problems for them, but more often I prefer to guide them to solving the exercises on their own. Often I find that students are unable to do the exercises because they do not know some of the basic definitions. For example, to answer a question such as "Why must a homogeneous linear system always be consistent?", you have to know what a homogeneous linear system is and what it means for a system to be consistent. Often a student will be completely baffled by a question, but, after we go over the definitions together, the student is then able to answer the question correctly.

The major results in the textbook are usually stated in the form of theorems. Many of the exercises test to see if students understand the theorems. In order to do these exercises it is essential for you be familiar with the theorems and understand what they say. Frequently students are stumped on an exercise but then are able to solve it after I refer them to the proper theorem. As the course progresses and students learn to study the theorems on their own, they find that they do not need to come in for help as often. Your success in learning linear algebra will depend on how well you master the definitions and theorems. To aid you in doing this, the study guide provides of list of the definitions and theorems for each section of the textbook.

The study guide is divided into seven chapters and each chapter is divided into sections corresponding to the sections in the text book. Each of the sections has four major components. These components are as follows:

1

- **Overview.** This subsection provides a brief overview of the topics covered in the section. Many students study the individual topics in a section but never put things together and see the bigger picture. The goal of the overview is summarize the section and put it into context. It is designed to help students see the significance of the material covered in the section.

- **Important Concepts.** This subsection lists the major definitions and terminology introduced in the section. You should study this list until you are familiar with all of the concepts. These terms should become part of your vocabulary for the course. You should not have to run and look them up every time they are used. While it is not necessary to memorize verbatim all of the definitions, there are some that are very basic. I do recommend that these be memorized and I often ask students in my classes to state them on exams. The following is a list of 14 definitions (sorted by chapter) that I consider to be the most important.

  Chapter 1

  scalar product of two vectors, matrix multiplication

  Chapter 3

  subspace of a vector space, linearly independent vectors, linearly dependent vectors, the span of a set of vectors, spanning set for a vector space, basis for a vector space, rank of a matrix.

  Chapter 5

  orthogonal vectors, inner product of two vectors, orthonormal vectors

  Chapter 6

  eigenvalue, eigenvector

- **Important Theorems and Results.** This subsection lists the theorems and important results that are included in the section. Note that in many sections not all of the important results are stated in the form of theorems. While it is not necessary to memorize the theorems, you should become familiar with what the theorems say. You should understand all theorems and important results well enough so that you are able to refer back to them at any future time during the course.

- **Exercises: Solutions and Hints.** This subsection provides solutions and hints for the exercises in the section. The hints are designed to get you thinking on the right track. This may be done by asking a question, referring you to a theorem in the book, or just telling you how to get started on the problem.

Steven J. Leon
sleon@umassd.edu

Chapter 1

Matrices and Systems of Equations

1 | SYSTEMS OF LINEAR EQUATIONS

Overview

This section provides an introduction to solving linear systems. The section concentrates on linear systems that have a unique solution. To master the section you should learn how to represent a linear system by an augmented matrix and how to use elementary row operations to reduce the matrix to strict triangular form. A system in this form can easily be solved using a technique called back substitution that successively determines the values of each of the unknown variables.

Important Concepts

1. **Solution to a linear system.** An n-tuple $(x_1, x_2, \ldots, x_n)$ is said to be a *solution* to a system of m linear equations in n unknowns if $(x_1, x_2, \ldots, x_n)$ satisfies all the equations of the system. The set of all solutions is referred to as the *solution set* for the system.

2. **Consistent system** A linear system of equations is said to be *consistent* if it has at least one solution.

3. **Inconsistent system.** A linear system of equations is said to be *inconsistent* if it does not have any solutions.

4. **Equivalent systems.** Two linear systems involving the same unknowns are said to be *equivalent* if they have the same solution set.

5. **Coefficient Matrix.** The *coefficient matrix* for a linear system is the two dimensional array of coefficients in a linear system. Each row of the coefficient matrix of a $m \times n$ linear system contains the n coefficients for the corresponding equation in the system.

6. **Augmented Matrix.** When an additional column corresponding to the right hand side of a linear system is attached to the coefficient matrix we refer to the resulting matrix as an *augmented matrix*. We can also augment a matrix by attaching more than one additional columns.

7. **Strict Triangular form.** A linear system is said to be in *strict triangular form* if the diagonal entries of its coefficient matrix are all nonzero and the entries below the diagonal are all equal to 0.

8. **Back Substitution.** *Back substitution* is a method for solving systems that are in triangular form. Basically you solve the last equation for x_n, substitute that value into the $n - 1$st equation, solve for x_{n-1}, substitute both values in the previous equation and so on.

9. **Elementary Row Operations.** These are the basic operations that we can perform to transform the augmented matrix into a simpler form so that the corresponding linear system will be easier to solve. The *elementary row operations* are:

    I. Interchange two rows.
    II. Multiply a row by a nonzero real number.
    III. Replace a row by its sum with a multiple of another row.

These row operations do not alter the solution to the system. After these operations have been performed the resulting augmented matrix will represent a linear system that is equivalent to the original system.

Exercises: Solutions and Hints

2. (d) **Solution.** The coefficient matrix for the system is

$$\begin{pmatrix} 1 & 1 & 1 & 1 & 1 \\ 0 & 2 & 1 & -2 & 1 \\ 0 & 0 & 4 & 1 & -2 \\ 0 & 0 & 0 & 1 & -3 \\ 0 & 0 & 0 & 0 & 2 \end{pmatrix}$$

5. Solution. The systems corresponding to the augmented matrices are:

(a) $3x_1 + 2x_2 = 8$
$\quad x_1 + 5x_2 = 7$

(c) $2x_1 + \quad x_2 + 4x_3 = -1$
$\quad 4x_1 - 2x_2 + 3x_3 = \quad 4$
$\quad 5x_1 + 2x_2 + 6x_2 = -1$

6. (d) **Solution.** The augmented matrix for this system is

$$\left(\begin{array}{ccc|c} 1 & 2 & -1 & 1 \\ 2 & -1 & 1 & 3 \\ -1 & 2 & 3 & 7 \end{array} \right)$$

To start we need to eliminate the last 2 entries in the first column of the matrix. To eliminate the (2,1) entry of the matrix we subtract 2 times the first row to the second row. In other words, we multiply the entries of the first row by 2

$$2 \left(\begin{array}{ccc|c} 1 & 2 & -1 & 1 \end{array} \right) = \left(\begin{array}{ccc|c} 2 & 4 & -2 & 2 \end{array} \right)$$

and then subtract the result from the second row.

$$\left(\begin{array}{ccc|c} 2 & -1 & 1 & 3 \\ 2 & 4 & -2 & 2 \\ \hline 0 & -5 & 3 & 1 \end{array} \right)$$

We use this result as our new second row. Thus the augmented matrix is now

$$\left(\begin{array}{rrr|r} 1 & 2 & -1 & 1 \\ 0 & -5 & 3 & 1 \\ -1 & 2 & 3 & 7 \end{array} \right)$$

To eliminate the $(3, 1)$ entry we add the first and third rows

$$\left(\begin{array}{rrr|r} 1 & 2 & -1 & 1 \\ -1 & 2 & 3 & 7 \\ \hline 0 & 4 & 2 & 8 \end{array} \right)$$

We use this result as our new third row. The augmented matrix is now in the form

$$\left(\begin{array}{rrr|r} 1 & 2 & -1 & 1 \\ 0 & -5 & 3 & 1 \\ 0 & 4 & 2 & 8 \end{array} \right)$$

Next we need to eliminate the $(2, 3)$ entry of the current matrix. We could do this by adding $4/5$ times the second row to the third, or, if we want to avoid using fractions, we could multiply the second row by 4 and the third row by 5 and add the results.

$$\begin{array}{rcl} 4(0 \ -5 \quad 3 \mid 1) & = & (0 \ -20 \quad 12 \mid 4) \\ 5(0 \quad 4 \quad 2 \mid 8) & = & \underline{(0 \quad 20 \quad 10 \mid 40)} \\ & = & (0 \quad 0 \quad 22 \mid 44) \end{array}$$

This is used as the new third row. The augmented matrix has now been reduced to triangular form.

$$\left(\begin{array}{rrr|r} 1 & 2 & -1 & 1 \\ 0 & -5 & 3 & 1 \\ 0 & 0 & 22 & 44 \end{array} \right)$$

We can now solve the triangular form system using back substitution. It follows form the third equation that $22x_3 = 44$ and hence $x_3 = 2$. Using this value in the second equation, $-5x_2 + 3x_3 = 1$, we get $-5x_2 + 6 = 1$, and hence $x_2 = 1$. Finally we plug in $x_2 = 1$ and $x_3 = 2$ into the first equation

$$x_1 + 2x_2 - x_3 = 1$$

and solve for x_1.

$$x_1 + 2 - 2 = 1$$

It follows that $x_1 = 1$ and hence the solution to the system is the ordered triple $(1, 1, 2)$.

9. **Hints.**

   (a) and (b)

   Subtract the first equation from the second. Now replace the second equation in the system by this new equation. Under what conditions will this new system have a unique solution?

   (c) If the geometric interpretation of an equation of the form $-mx_1 + x_2 = b$ is not immediately apparent to you, then think of x_1 as your independent variable x and x_2 as your dependent variable y. If you plug x and y into the equation and solve for y then you get

   $$y = mx + b$$

   This equation should be familiar to you from your elementary algebra courses. What is the geometrical interpretation of the equation and of the quantities m and b?

10. **Hint.** When the numbers on the right hand side of the equations are both 0 there is an obvious choice x_1 and x_2 that will make both left hand sides equal to 0.

11. **Hint.** A linear equation in 3 unknowns represents a plane in 3-space. The solution sets for a 3×3 linear system would be the set of all points that lie on all three planes. What are the possibilities for the ways that three planes can intersect?

2 | ROW ECHELON FORM

Overview

In this section we generalize the method of Section 1 to apply to systems that do not have a unique solution and systems where the number of equations does not necessarily equal the number of unknowns. The method presented is known as *Gaussian elimination*. The general strategy can be summarized as follows.

1. Use row operations to eliminate entries below the diagonal of the augmented matrix. If the system can be transformed to one whose nonzero rows are in strict triangular form, then there is a unique solution. It can be found using back substitution.

2. If the system cannot be transformed to strict triangular form, then use row operations to reduce the augmented matrix to row echelon form.

3. If the row echelon form of the augmented matrix has a row of the form $(0 \ 0 \ \dots \ 0 \,|\, 1)$, then the system is inconsistent.

4. If the row echelon form of the augmented matrix does not have a row of the form $(0 \ 0 \ \dots \ 0 \,|\, 1)$, then the system is consistent. If the system is consistent

and the row echelon form is not in strict triangular form then there must be at least one free variable. In this case continue the elimination process until the system is in reduced row echelon form. In this form it is easy to solve for the lead variables in terms of the free variables. You can then represent the general solution by an n-tuple whose entries involve only constants and free variables. Since the free variables can be assigned any values, the system will have infinitely many solutions.

Important Concepts

1. **Row Echelon Form.** A matrix is said to be in *row echelon form* if
   (i) The first nonzero entry in each row is 1.
   (ii) If row k does not consist entirely of zeros, the number of leading zero entries in row $k + 1$ is greater than the number of leading zero entries in row k.
   (iii) If there are rows whose entries are all zero, they are below the rows having nonzero entries.

2. **Reduced Row Echelon Form.** A matrix is said to be in *reduced row echelon form* if:
   (i) The matrix is in row echelon form.
   (ii) The first nonzero entry in each row is the only nonzero entry in its column.

3. **Gaussian Elimination.** The process of using row operations I, II, and III to transform a linear system into one whose augmented matrix is in row echelon form is called *Gaussian elimination.*

4. **Underdetermined Systems.** A system of m linear equations in n unknowns is said to be *underdetermined* if there are fewer equations than unknowns $(m < n)$.

5. **Overdetermined Systems.** A linear system is said to be *overdetermined* if there are more equations than unknowns.

6. **Homogeneous Systems.** A system of linear equations is said to be *homogeneous* if the constants on the right-hand side are all zero.

Important Theorem

▶ THEOREM 1.2.1. *An $m \times n$ homogeneous system of linear equations has a nontrivial solution if $n > m$.*

   We will make use of this theorem often in the book. The result is nice in that it is useful in the course and it is easy to see why the theorem is true. Indeed, if a linear system is homogeneous, then it is consistent since $(0, \ldots, 0)$ is a solution. Since there are more unknowns than equations, the echelon form of the system must involve a free variable. A consistent system with a free variable will have infinitely many solutions.

Exercises: Solutions and Hints

5. (j) **Solution.** To solve the system we first compute the row echelon form of the augmented matrix. We begin by eliminating the (2,1) and (3,1) entries.

$$\begin{pmatrix} 1 & 2 & -3 & 1 & | & 1 \\ -1 & -1 & 4 & -1 & | & 6 \\ -2 & -4 & 7 & -1 & | & 1 \end{pmatrix} \rightarrow \begin{pmatrix} 1 & 2 & -3 & 1 & | & 1 \\ 0 & 1 & 1 & 0 & | & 7 \\ -2 & -4 & 7 & -1 & | & 1 \end{pmatrix}$$

$$\rightarrow \begin{pmatrix} 1 & 2 & -3 & 1 & | & 1 \\ 0 & 1 & 1 & 0 & | & 7 \\ 0 & 0 & 1 & 1 & | & 3 \end{pmatrix}$$

The (3,2) coefficient also was eliminated and since the lead coefficients in each row are all 1's the matrix is in row echelon form. Since the system is consistent and there is a free variable x_4, we proceed on to computing the reduced row echelon form. To do this we first eliminate the entries in the third column that are above the diagonal

$$\begin{pmatrix} 1 & 2 & -3 & 1 & | & 1 \\ 0 & 1 & 1 & 0 & | & 7 \\ 0 & 0 & 1 & 1 & | & 3 \end{pmatrix} \rightarrow \begin{pmatrix} 1 & 2 & 0 & 4 & | & 10 \\ 0 & 1 & 1 & 0 & | & 7 \\ 0 & 0 & 1 & 1 & | & 3 \end{pmatrix}$$

$$\rightarrow \begin{pmatrix} 1 & 2 & 0 & 4 & | & 10 \\ 0 & 1 & 0 & -1 & | & 4 \\ 0 & 0 & 1 & 1 & | & 3 \end{pmatrix}$$

and then we eliminate the (1,2) entry of the matrix.

$$\begin{pmatrix} 1 & 2 & 0 & 4 & | & 10 \\ 0 & 1 & 0 & -1 & | & 4 \\ 0 & 0 & 1 & 1 & | & 3 \end{pmatrix} \rightarrow \begin{pmatrix} 1 & 0 & 0 & 6 & | & 2 \\ 0 & 1 & 0 & -1 & | & 4 \\ 0 & 0 & 1 & 1 & | & 3 \end{pmatrix}$$

If we set the free variable $x_4 = \alpha$, we can then solve the system for the lead variables as functions of α. We get

$$x_1 = 2 - 6\alpha, \quad x_2 = 4 + \alpha, \quad x_3 = 3 - \alpha$$

For each possible value of α we get a solution. It follows that the solution set consists of all 4-tuples of the form $(2 - 5\alpha, 4 + \alpha, 3 - \alpha, \alpha)$.

7. Hint. The hint for Exercise 11 in the previous section also applies to this exercise.

15. Hint. Plug in $x_1 = \alpha c_1$ and $x_2 = \alpha c_2$ into the system to see if they work.

---

3 │ MATRIX ALGEBRA

Overview

This section introduces the standard notation that is used for matrices and vectors and the standard algebraic operations that are used for matrices.

Important Concepts

1. **Matrix and vector notation.** Matrices are represented by capital letters, such as A, B, C, and column vectors are represented by boldface lowercase letters, such as, $\mathbf{x}, \mathbf{y}, \mathbf{z}$. The (i, j) entry of a matrix A is denoted by a_{ij}. The jth column of A is represented by $\mathbf{a}_j$ and the ith row is denoted $\mathbf{a}(i, :)$.

2. **Scalars.** The entries of a matrix are referred to as *scalars*. Generally a scalar is either a real or a complex number. In chapters 1–5 the term scalar always refers to a real number. Complex scalars are not introduced until Chapter 6.

3. **R^n.** The set of all column vectors with n entries, i.e, all $n \times 1$ matrices of real numbers, is called *Euclidean n-space* and is denoted R^n.

4. **Scalar multiplication.** If A is a matrix and α is a scalar, then αA is the matrix whose (i, j) entry is αa_{ij}.

5. **Matrix Addition.** If $A = (a_{ij})$ and $B = (b_{ij})$ are both $m \times n$ matrices, then the *sum* $A + B$ is the $m \times n$ matrix whose (i, j) entry is $a_{ij} + b_{ij}$ for each ordered pair (i, j).

6. **Matrix multiplication.** If $A = (a_{ij})$ is an $m \times n$ matrix and $B = (b_{ij})$ is an $n \times r$ matrix, then the *product* $AB = C = (c_{ij})$ is the $m \times r$ matrix whose entries are defined by

$$c_{ij} = \mathbf{a}(i, :)\mathbf{b}_j = \sum_{k=1}^{n} a_{ik}b_{kj}$$

7. **Linear combination.** If $\mathbf{a}_1, \mathbf{a}_2, \ldots, \mathbf{a}_n$ are vectors in R^m and $c_1, c_2, \ldots, c_n$ are scalars, then a sum of the form

$$c_1\mathbf{a}_1 + c_2\mathbf{a}_2 + \cdots + c_n\mathbf{a}_n$$

is said to be a *linear combination* of the vectors $\mathbf{a}_1, \mathbf{a}_2, \ldots, \mathbf{a}_n$.

8. **Identity matrix.** The $n \times n$ *identity matrix* is the matrix $I = (\delta_{ij})$, where

$$\delta_{ij} = \begin{cases} 1 & \text{if } i = j \\ 0 & \text{if } i \neq j \end{cases}$$

The standard notation for the jth column vector of I is $\mathbf{e}_j$ rather than the usual $\mathbf{i}_j$.

9. **Nonsingular matrix.** An $n \times n$ matrix A is said to be *nonsingular* or *invertible* if there exists a matrix B such that $AB = BA = I$. The matrix B is said to be a *multiplicative inverse* of A. If a matrix A has a multiplicative inverse then it is unique. The standard notation for the multiplicative inverse of A is A^{-1}.

10. **Singular matrix.** An $n \times n$ matrix is said to be *singular* if it does not have a multiplicative inverse.

11. **Transpose of a matrix.** The *transpose* of an $m \times n$ matrix A is the $n \times m$ matrix B defined by
$$b_{ji} = a_{ij}$$

12. **Symmetric matrix** An $n \times n$ matrix A is said to be *symmetric* if $A^T = A$.

Important Theorems and Results

1. If A is an $m \times n$ matrix and $\mathbf{x}$ is a vector in R^n, then
$$A\mathbf{x} = x_1\mathbf{a}_1 + x_2\mathbf{a}_2 + \cdots + x_n\mathbf{a}_n$$

2. ▶ THEOREM 1.3.1 **(Consistency Theorem for Linear Systems)** *A linear system* $A\mathbf{x} = \mathbf{b}$ *is consistent if and only if* $\mathbf{b}$ *can be written as a linear combination of the column vectors of* A.

   ▶ THEOREM 1.3.3. *If A and B are nonsingular $n \times n$ matrices, then AB is also nonsingular and $(AB)^{-1} = B^{-1}A^{-1}$.*

Exercises: Solutions and Hints

1. (g) **Solution.**
$$A^T B^T = \begin{pmatrix} 5 & -10 & 15 \\ 5 & -1 & 4 \\ 8 & -9 & 6 \end{pmatrix}$$

5. **Solution.**

(c) $A^T = \begin{pmatrix} 3 & 1 & 2 \\ 4 & 1 & 7 \end{pmatrix}$

$$(A^T)^T = \begin{pmatrix} 3 & 1 & 2 \\ 4 & 1 & 7 \end{pmatrix}^T = \begin{pmatrix} 3 & 4 \\ 1 & 1 \\ 2 & 7 \end{pmatrix} = A$$

6. (b) **Solution**.

$$3(A+B) = 3\begin{bmatrix} 5 & 4 & 6 \\ 0 & 5 & 1 \end{bmatrix} = \begin{bmatrix} 15 & 12 & 18 \\ 0 & 15 & 3 \end{bmatrix}$$

$$3A + 3B = \begin{bmatrix} 12 & 3 & 18 \\ 6 & 9 & 15 \end{bmatrix} + \begin{bmatrix} 3 & 9 & 0 \\ -6 & 6 & -12 \end{bmatrix} = \begin{bmatrix} 15 & 12 & 18 \\ 0 & 15 & 3 \end{bmatrix}$$

7. (a) **Solution**.

$$3(AB) = 3\begin{bmatrix} 5 & 14 \\ 15 & 42 \\ 0 & 16 \end{bmatrix} = \begin{bmatrix} 15 & 42 \\ 45 & 126 \\ 0 & 48 \end{bmatrix}$$

$$(3A)B = \begin{bmatrix} 6 & 3 \\ 18 & 9 \\ -6 & 12 \end{bmatrix} \begin{bmatrix} 2 & 4 \\ 1 & 6 \end{bmatrix} = \begin{bmatrix} 15 & 42 \\ 45 & 126 \\ 0 & 48 \end{bmatrix}$$

$$A(3B) = \begin{bmatrix} 2 & 1 \\ 6 & 3 \\ -2 & 4 \end{bmatrix} \begin{bmatrix} 6 & 12 \\ 3 & 18 \end{bmatrix} = \begin{bmatrix} 15 & 42 \\ 45 & 126 \\ 0 & 48 \end{bmatrix}$$

8. (b) **Solution**.

$$(AB)C = \begin{bmatrix} -4 & 18 \\ -2 & 13 \end{bmatrix} \begin{bmatrix} 3 & 1 \\ 2 & 1 \end{bmatrix} = \begin{bmatrix} 24 & 14 \\ 20 & 11 \end{bmatrix}$$

$$A(BC) = \begin{bmatrix} 2 & 4 \\ 1 & 3 \end{bmatrix} \begin{bmatrix} -4 & -1 \\ 8 & 4 \end{bmatrix} = \begin{bmatrix} 24 & 14 \\ 20 & 11 \end{bmatrix}$$

9. Solution. Let

$$D = (AB)C = \begin{bmatrix} a_{11}b_{11} + a_{12}b_{21} & a_{11}b_{12} + a_{12}b_{22} \\ a_{21}b_{11} + a_{22}b_{21} & a_{21}b_{12} + a_{22}b_{22} \end{bmatrix} \begin{bmatrix} c_{11} & c_{12} \\ c_{21} & c_{22} \end{bmatrix}$$

It follows that

$$\begin{aligned} d_{11} &= (a_{11}b_{11} + a_{12}b_{21})c_{11} + (a_{11}b_{12} + a_{12}b_{22})c_{21} \\ &= a_{11}b_{11}c_{11} + a_{12}b_{21}c_{11} + a_{11}b_{12}c_{21} + a_{12}b_{22}c_{21} \end{aligned}$$

$$d_{12} = (a_{11}b_{11} + a_{12}b_{21})c_{12} + (a_{11}b_{12} + a_{12}b_{22})c_{22}$$
$$= a_{11}b_{11}c_{12} + a_{12}b_{21}c_{12} + a_{11}b_{12}c_{22} + a_{12}b_{22}c_{22}$$
$$d_{21} = (a_{21}b_{11} + a_{22}b_{21})c_{11} + (a_{21}b_{12} + a_{22}b_{22})c_{21}$$
$$= a_{21}b_{11}c_{11} + a_{22}b_{21}c_{11} + a_{21}b_{12}c_{21} + a_{22}b_{22}c_{21}$$
$$d_{22} = (a_{21}b_{11} + a_{22}b_{21})c_{12} + (a_{21}b_{12} + a_{22}b_{22})c_{22}$$
$$= a_{21}b_{11}c_{12} + a_{22}b_{21}c_{12} + a_{21}b_{12}c_{22} + a_{22}b_{22}c_{22}$$

If we set

$$E = A(BC) = \begin{pmatrix} a_{11} & a_{12} \\ a_{21} & a_{22} \end{pmatrix} \begin{pmatrix} b_{11}c_{11} + b_{12}c_{21} & b_{11}c_{12} + b_{12}c_{22} \\ b_{21}c_{11} + b_{22}c_{21} & b_{21}c_{12} + b_{22}c_{22} \end{pmatrix}$$

then it follows that

$$e_{11} = a_{11}(b_{11}c_{11} + b_{12}c_{21}) + a_{12}(b_{21}c_{11} + b_{22}c_{21})$$
$$= a_{11}b_{11}c_{11} + a_{11}b_{12}c_{21} + a_{12}b_{21}c_{11} + a_{12}b_{22}c_{21}$$
$$e_{12} = a_{11}(b_{11}c_{12} + b_{12}c_{22}) + a_{12}(b_{21}c_{12} + b_{22}c_{22})$$
$$= a_{11}b_{11}c_{12} + a_{11}b_{12}c_{22} + a_{12}b_{21}c_{12} + a_{12}b_{22}c_{22}$$
$$e_{21} = a_{21}(b_{11}c_{11} + b_{12}c_{21}) + a_{22}(b_{21}c_{11} + b_{22}c_{21})$$
$$= a_{21}b_{11}c_{11} + a_{21}b_{12}c_{21} + a_{22}b_{21}c_{11} + a_{22}b_{22}c_{21}$$
$$e_{22} = a_{21}(b_{11}c_{12} + b_{12}c_{22}) + a_{22}(b_{21}c_{12} + b_{22}c_{22})$$
$$= a_{21}b_{11}c_{12} + a_{21}b_{12}c_{22} + a_{22}b_{21}c_{12} + a_{22}b_{22}c_{22}$$

Thus

$$d_{11} = e_{11} \qquad d_{12} = e_{12} \qquad d_{21} = e_{21} \qquad d_{22} = e_{22}$$

and hence

$$(AB)C = D = E = A(BC)$$

13. (a) **Solution.** $\mathbf{b} = 2\mathbf{a}_1 + \mathbf{a}_2$.

15. Solution. If $d = a_{11}a_{22} - a_{21}a_{12} \neq 0$ then

$$\frac{1}{d} \begin{pmatrix} a_{22} & -a_{12} \\ -a_{21} & a_{11} \end{pmatrix} \begin{pmatrix} a_{11} & a_{12} \\ a_{21} & a_{22} \end{pmatrix} =$$

$$\frac{1}{d} \begin{pmatrix} a_{22}a_{11} - a_{12}a_{21} & 0 \\ 0 & -a_{21}a_{12} + a_{11}a_{22} \end{pmatrix} = I$$

and

$$\begin{pmatrix} a_{11} & a_{12} \\ a_{21} & a_{22} \end{pmatrix} \begin{bmatrix} \frac{1}{d} \begin{pmatrix} a_{22} & -a_{12} \\ -a_{21} & a_{11} \end{pmatrix} \end{bmatrix} =$$

$$\frac{1}{d} \begin{pmatrix} a_{11}a_{12} - a_{12}a_{21} & 0 \\ & \\ 0 & -a_{21}a_{12} + a_{22}a_{11} \end{pmatrix} = I$$

Therefore

$$\frac{1}{d} \begin{pmatrix} a_{22} & -a_{12} \\ -a_{21} & a_{11} \end{pmatrix} = A^{-1}$$

16. **Hint.** You should be able to show this result directly from the definition of a nonsingular matrix.

18. **Hint.** If A were nonsingular and $A\mathbf{x} = A\mathbf{y}$, what could you conclude about the vectors $\mathbf{x}$ and $\mathbf{y}$?

19. **Solution.** For $m = 1$,

$$(A^1)^{-1} = A^{-1} = (A^{-1})^1$$

Assume the result holds in the case $m = k$, that is,

$$(A^k)^{-1} = (A^{-1})^k$$

It follows that

$$(A^{-1})^{k+1} A^{k+1} = A^{-1}(A^{-1})^k A^k A = A^{-1} A = I$$

and

$$A^{k+1}(A^{-1})^{k+1} = A A^k (A^{-1})^k A^{-1} = A A^{-1} = I$$

Therefore

$$(A^{-1})^{k+1} = (A^{k+1})^{-1}$$

and the result follows by mathematical induction.

20. **Hint**. In general

$$BA + AB \neq 2AB \qquad \text{and} \qquad BA - AB \neq O$$

22. **Hint.** There are many possible choices for A and B. Make up a nonzero matrix A whose first and second rows are the same and then see if you can come up with a matrix B such that $AB = O$.

23. **Hint.** Make use of the result from Exercise 22.

24. **Solution.** A 2×2 symmetric matrix is one of the form

$$A = \begin{pmatrix} a & b \\ b & c \end{pmatrix}$$

If A is of this form and $A^2 = O$, what would that imply about the scalars a, b and c?

25. Hint. Try some examples.

27. Solution. Let A and B be symmetric $n \times n$ matrices. If $(AB)^T = AB$ then

$$BA = B^T A^T = (AB)^T = AB$$

Conversely if $BA = AB$ then

$$(AB)^T = B^T A^T = BA = AB$$

29. (b) **Hint.** How is the matrix $\frac{1}{2}B + \frac{1}{2}C$ related to A?

31. Solution. The search vector is $\mathbf{x} = (1, 0, 1, 0, 1, 0)^T$. The search result is given by the vector

$$\mathbf{y} = A^T\mathbf{x} = (1, 2, 2, 1, 1, 2, 1)^T$$

The ith entry of $\mathbf{y}$ is equal to the number of search words in the title of the ith book.

34. Solution. If $\alpha = a_{21}/a_{11}$, then

$$\begin{pmatrix} 1 & 0 \\ \alpha & 1 \end{pmatrix} \begin{pmatrix} a_{11} & a_{12} \\ 0 & b \end{pmatrix} = \begin{pmatrix} a_{11} & a_{12} \\ \alpha a_{11} & \alpha a_{12} + b \end{pmatrix} = \begin{pmatrix} a_{11} & a_{12} \\ a_{21} & \alpha a_{12} + b \end{pmatrix}$$

The product will equal A provided

$$\alpha a_{12} + b = a_{22}$$

Thus we must choose

$$b = a_{22} - \alpha a_{12} = a_{22} - \frac{a_{21}a_{12}}{a_{11}}$$

4 | ELEMENTARY MATRICES

Overview

Elementary matrices are special matrices that allow us to treat row operations as matrix multiplications. This allows us to view the elimination process in terms of a factorization of the coefficient matrix. Furthermore, using elementary matrices we are able to derive a method for computing the inverse of a matrix. This section also includes an important theorem that characterizes when a matrix is nonsingular.

Important Concepts

1. **Elementary matrices.** An *elementary matrix* is a matrix that is formed by performing exactly one elementary row operation on the identity matrix.

2. **Row equivalent matrices.** A matrix B is *row equivalent* to A if there exists a finite sequence $E_1, E_2, \ldots, E_k$ of elementary matrices such that

$$B = E_k E_{k-1} \cdots E_1 A$$

3. **Triangular matrices.** An $n \times n$ matrix A is said to be *upper triangular* if $a_{ij} = 0$ for $i > j$ and *lower triangular* if $a_{ij} = 0$ for $i < j$. Also, A is said to be *triangular* if it is either upper triangular or lower triangular.

   Note that a triangular matrix may have 0's on the diagonal. However, for a linear system $A\mathbf{x} = \mathbf{b}$ to be in strict triangular form, the coefficient matrix A must be triangular with nonzero diagonal entries.

4. **Diagonal matrices.** An $n \times n$ matrix A is *diagonal* if $a_{ij} = 0$ whenever $i \neq j$.

   Note that a diagonal matrix is both upper and lower triangular.

5. *LU* **factorization.** If an $n \times n$ matrix A can be reduced to upper triangular form using only row operation III, then it possible to factor A into a product LU where L is lower triangular with 1's on its diagonal and U is an upper triangular matrix.

Important Theorems

▶ THEOREM 1.4.1. *If E is an elementary matrix, then E is nonsingular and E^{-1} is an elementary matrix of the same type.*

▶ THEOREM 1.4.2 **(Equivalent Conditions for Nonsingularity)** *Let A be an $n \times n$ matrix. The following are equivalent:*

(a) *A is nonsingular.*
(b) *$A\mathbf{x} = \mathbf{0}$ has only the trivial solution $\mathbf{x} = \mathbf{0}$.*
(c) *A is row equivalent to I.*

▶ COROLLARY 1.4.3. *The system of n linear equations in n unknowns $A\mathbf{x} = \mathbf{b}$ has a unique solution if and only if A is nonsingular.*

Exercises: Solutions and Hints

2. (c) **Solution.** The given matrix is an elementary matrix of type III. The matrix can be transformed back to the identity by subtracting 5 times the first row from the third row. The inverse is the elementary matrix of type III that does this row operation, i.e., the matrix

$$\begin{bmatrix} 1 & 0 & 0 \\ 0 & 1 & 0 \\ -5 & 0 & 1 \end{bmatrix}$$

5. (c) **Hint.** See if the definition of row equivalent matrices applies.

7. **Hint.** For part (b) use row operations to transform A to reduced row echelon form and keep track of the elementary matrices involved. Use the result from part (b) to obtain the factorization for part (a).

8. (d) **Hint and Partial Solution.** The matrix can be reduced to an upper triangular form U as follows

$$\begin{pmatrix} -2 & 1 & 2 \\ 4 & 1 & -2 \\ -6 & -3 & -4 \end{pmatrix} \rightarrow \begin{pmatrix} -2 & 1 & 2 \\ 0 & 3 & 2 \\ 0 & -6 & -10 \end{pmatrix}$$

$$\rightarrow \begin{pmatrix} -2 & 1 & 2 \\ 0 & 3 & 2 \\ 0 & 0 & -6 \end{pmatrix}$$

In the first step, what multiples of the first row were subtracted from the second and third rows? In the second step of the process, what multiple of the second row was subtracted from the third? Use your answers to these questions to construct L. Check to see if your factorization works by computing LU.

9. (a) **Hint.** According to the definition $B = A^{-1}$ if and only if

$$AB = BA = I$$

so check to see if the given inverse matrix has this property.

(b) **Hint.** If $A\mathbf{x} = \mathbf{b}$ and A is nonsingular, then $\mathbf{x} = A^{-1}\mathbf{b}$.

12. (d) **Hint.** If $XA + C = X$, then

$$\begin{aligned} XA - XI &= -C \\ X(A - I) &= -C \\ X &= -C(A - I)^{-1} \end{aligned}$$

13. **Hint.** For each of the two questions try some examples using elementary matrices of all three types. If the examples all seem to work, see if you can explain why the answer to the question is yes. If you find an example that doesn't satisfy the property in question, then the example shows that the property doesn't hold in general.

14. **Solution.** If $T = UR$, then

$$t_{ij} = \sum_{k=1}^{n} u_{ik} r_{kj}$$

Since U and R are upper triangular

$$\begin{aligned} u_{i1} &= u_{i2} = \cdots = u_{i,i-1} = 0 \\ r_{j+1,j} &= r_{j+2,j} = \cdots - r_{nj} = 0 \end{aligned}$$

If $i > j$, then

$$t_{ij} = \sum_{k=1}^{j} u_{ik} r_{kj} + \sum_{k=j+1}^{n} u_{ik} r_{kj}$$

$$= \sum_{k=1}^{j} 0\, r_{kj} + \sum_{k=j+1}^{n} u_{ik} 0$$

$$= 0$$

Therefore T is upper triangular.

If $i = j$, then

$$t_{jj} = t_{ij} = \sum_{k=1}^{i-1} u_{ik} r_{kj} + u_{jj} r_{jj} + \sum_{k=j+1}^{n} u_{ik} r_{kj}$$

$$= \sum_{k=1}^{i-1} 0\, r_{kj} + u_{jj} r_{jj} + \sum_{k=j+1}^{n} u_{ik} 0$$

$$= u_{jj} r_{jj}$$

Therefore

$$t_{jj} = u_{jj} r_{jj} \qquad j = 1, \ldots, n$$

15. **Hint.** The given information should be sufficient for you to be able to determine two solutions. If you can find two solutions, then what does that tell you about the total number of solutions to the linear system.

16. **Hint.** Rewrite the equation in the same form as the equation in Exercise 15.

17. **Hint.** Look at $C\mathbf{x}_0$.

18. **Hint.** The hint given in the book should be sufficient, but here is a second hint. The result can be proved using condition (b) of Theorem 1.4.2.

19. (a) **Hint.** To explain why U must be nonsingular make use of condition (c) in Theorem 1.4.2.
    (b) **Hint.** The same row operations that were used to reduce U to the identity matrix will transform I into U^{-1}. What effect to these operations have when you apply them I?

20. **Solution.** Since A is nonsingular it is row equivalent to I. Hence there exist elementary matrices $E_1, E_2, \ldots, E_k$ such that

$$E_k \cdots E_1 A = I$$

It follows that

$$A^{-1} = E_k \cdots E_1$$

and

$$E_k \cdots E_1 B = A^{-1} B = C$$

The same row operations that reduce A to I, will transform B to C. Therefore the reduced row echelon form of $(A \mid B)$ will be $(I \mid C)$.

22. Hint. To show that A^{-1} is a symmetric matrix you need to show that $(A^{-1})^T = A^{-1}$. Going back to the definition of A^{-1}, to show $(A^{-1})^T = A^{-1}$ you must show that

$$(A^{-1})^T A = I \quad \text{and} \quad A(A^{-1})^T = I$$

23. Hint. Write out what it means for A to be row equivalent to B and then make use of Theorem 1.4.1.

24. (a) **Hint.** If A is row equivalent to B, then there exist elementary matrices $E_1, E_2, \ldots, E_k$ such that

$$A = E_k E_{k-1} \cdots E_1 B$$

Since B is row equivalent to C, there exist elementary matrices $H_1, H_2, \ldots, H_j$ such that

$$B = H_j H_{j-1} \cdots H_1 C$$

Use these equations to express A as the product of a finite sequence of elementary matrices times C.

(b) **Hint.** Make use of Theorem 1.4.2.

26. Hint. You need to show two things

(i) If B is row equivalent to A, then $B = MA$, where M is nonsingular.

(ii) If $B = MA$, where M is nonsingular, then B is row equivalent to A.

Make use of Theorems 1.3.3 and 1.4.1 to show (i). To show (ii), make use of Theorem 1.4.2.

27. (b) **Hint.** Show that if $V\mathbf{c} = \mathbf{0}$, then the polynomial

$$p(x) = c_1 + c_2 x + \cdots + c_{n+1} x^n$$

has $n + 1$ roots? What can you conclude about $p(x)$ if you know it has degree n and it has $n + 1$ roots?

---

5 | PARTITIONED MATRICES

Overview

This section of the book shows how to partition matrices into submatrices or blocks. If the partitioning is done properly, then it is possible to express the product of two partitioned matrices A and B as a partitioned matrix C. In particular it is important to know when the block multiplication is possible and how to compute the blocks of C in terms of the blocks of A and B.

Important Concepts

1. **Partitioned matrices.** A matrix A can be *partitioned* into an array of smaller matrices by drawing horizontal lines between the rows and vertical lines between the columns. The smaller matrices are often referred to as *blocks*.

2. Block multiplication. If $C = AB$, where A and B are partitioned matrices and the column partition of A matches up with the row partition of B, then the block multiplication of A times B can be carried out in the same way that we do ordinary matrix multiplication, that is, the (i, j) block of C is computed by setting

$$C_{ij} = \sum_{k=1}^{n} A_{ik} B_{kj}$$

For example, if

$$A = \begin{pmatrix} A_{11} & A_{12} & A_{13} \\ A_{21} & A_{22} & A_{23} \end{pmatrix} = \left(\begin{array}{cc|cc|c} 1 & 1 & 2 & 2 & 3 \\ 4 & 4 & 5 & 5 & 6 \\ \hline 4 & 4 & 5 & 5 & 6 \end{array}\right)$$

and

$$B = \left(\begin{array}{cc|cc} 1 & 1 & 2 & 2 \\ 1 & 1 & 2 & 2 \\ \hline 3 & 3 & 4 & 4 \\ 3 & 3 & 4 & 4 \\ \hline 5 & 5 & 6 & 6 \end{array}\right)$$

and the product $C = AB$ is determined using block multiplication, then the $(2, 1)$ block of C is computed as follows:

$$C_{21} = A_{21}B_{11} + A_{22}B_{21} + A_{23}B_{31}$$

$$= \begin{pmatrix} 4 & 4 \\ 4 & 4 \end{pmatrix} \begin{pmatrix} 1 & 1 \\ 1 & 1 \end{pmatrix} + \begin{pmatrix} 5 & 5 \\ 5 & 5 \end{pmatrix} \begin{pmatrix} 3 & 3 \\ 3 & 3 \end{pmatrix} + \begin{pmatrix} 6 \\ 6 \end{pmatrix} \begin{pmatrix} 5 & 5 \end{pmatrix}$$

3. Scalar product – inner product. The matrix product $\mathbf{x}^T\mathbf{y}$ is the product of a row vector (a $1 \times n$ matrix) times a column vector (an $n \times 1$ matrix). The result will be a 1×1 matrix or simply a scalar.

$$\mathbf{x}^T\mathbf{y} = (x_1, x_2, \ldots, x_n) \begin{pmatrix} y_1 \\ y_2 \\ \vdots \\ y_n \end{pmatrix} = x_1 y_1 + x_2 y_2 + \cdots + x_n y_n$$

This type of product is referred to as a *scalar product* or an *inner product*.

4. **Outer product.** The matrix product $\mathbf{x}\mathbf{y}^T$ is the product of an $n \times 1$ matrix times an $1 \times n$ matrix. The result is a full $n \times n$ matrix.

$$\mathbf{x}\mathbf{y}^T = \begin{pmatrix} x_1 \\ x_2 \\ \vdots \\ x_n \end{pmatrix} (y_1, y_2, \ldots, y_n) = \begin{pmatrix} x_1 y_1 & x_1 y_2 & \cdots & x_1 y_n \\ x_2 y_1 & x_2 y_2 & \cdots & x_2 y_n \\ \vdots & & & \\ x_n y_1 & x_n y_2 & \cdots & x_n y_n \end{pmatrix}$$

The product $\mathbf{x}\mathbf{y}^T$ is referred to as the *outer product* of $\mathbf{x}$ and $\mathbf{y}$. The outer product matrix has special structure in that each of its rows is a multiple of $\mathbf{y}^T$ and each of its column vectors is a multiple of $\mathbf{x}$.

5. **Outer product expansion.** If we partition X into columns and Y^T into to rows and perform the block multiplication we see that XY^T can be represented as a sum of outer products of vectors.

$$XY^T = (\mathbf{x}_1, \mathbf{x}_2, \ldots, \mathbf{x}_n) \begin{pmatrix} \mathbf{y}_1^T \\ \mathbf{y}_2^T \\ \vdots \\ \mathbf{y}_n^T \end{pmatrix} = \mathbf{x}_1 \mathbf{y}_1^T + \mathbf{x}_2 \mathbf{y}_2^T + \cdots + \mathbf{x}_n \mathbf{y}_n^T$$

This representation is referred to as an *outer product expansion.*

Exercises: Solutions and Hints

2. **Hint.** Partition A^T into row vectors and A into column vectors and perform the block multiplication.

5. (c) **Solution.** Let A be the matrix on left and let A_{11}, A_{12}, A_{21}, A_{22} denote the blocks of A.

$$A = \begin{pmatrix} \frac{3}{5} & -\frac{4}{5} & 0 & 0 \\ \frac{4}{5} & \frac{3}{5} & 0 & 0 \\ \hline 0 & 0 & 1 & 0 \end{pmatrix} = \begin{pmatrix} A_{11} & A_{12} \\ A_{21} & A_{22} \end{pmatrix} = \begin{pmatrix} A_{11} & O \\ \mathbf{0}^T & \mathbf{e}_1^T \end{pmatrix}$$

where $\mathbf{e}_1$ is the first column vector of the 2×2 identity matrix. The matrix on the right is just

$$A^T = \begin{pmatrix} A_{11}^T & A_{21}^T \\ A_{12}^T & A_{22}^T \end{pmatrix} = \begin{pmatrix} A_{11}^T & \mathbf{0} \\ O & \mathbf{e}_1 \end{pmatrix}$$

Note that

$$A_{11}A_{11}^T = \begin{pmatrix} \frac{3}{5} & -\frac{4}{5} \\ \\ \frac{4}{5} & \frac{3}{5} \end{pmatrix} \begin{pmatrix} \frac{3}{5} & \frac{4}{5} \\ \\ -\frac{4}{5} & \frac{3}{5} \end{pmatrix} = \begin{pmatrix} 1 & 0 \\ 0 & 1 \end{pmatrix} = I$$

and hence block multiplication of AA^T is

$$\begin{pmatrix} A_{11} & A_{12} \\ \\ A_{21} & A_{22} \end{pmatrix} \begin{pmatrix} A_{11}^T & A_{21}^T \\ \\ A_{12}^T & A_{22}^T \end{pmatrix} = \begin{pmatrix} A_{11} & O \\ \\ \mathbf{0}^T & \mathbf{e}_1^T \end{pmatrix} \begin{pmatrix} A_{11}^T & \mathbf{0} \\ \\ O & \mathbf{e}_1 \end{pmatrix}$$

$$= \begin{pmatrix} A_{11}A_{11}^T + OO^T & A_{11}\mathbf{0} + O\mathbf{e}_1 \\ \\ \mathbf{0}^T A_{11}^T + \mathbf{e}_1^T O & \mathbf{0}^T\mathbf{0} + \mathbf{e}_1^T\mathbf{e}_1 \end{pmatrix}$$

$$= \begin{pmatrix} I & \mathbf{0} \\ \\ \mathbf{0}^T & 1 \end{pmatrix}$$

We end up with the 3×3 identity matrix.

6. (a) **Solution.**

$$XY^T = \mathbf{x}_1\mathbf{y}_1^T + \mathbf{x}_2\mathbf{y}_2^T + \mathbf{x}_3\mathbf{y}_3^T$$

$$= \begin{pmatrix} 2 \\ 4 \end{pmatrix} \begin{pmatrix} 1 & 2 \end{pmatrix} + \begin{pmatrix} 1 \\ 2 \end{pmatrix} \begin{pmatrix} 2 & 3 \end{pmatrix} + \begin{pmatrix} 5 \\ 3 \end{pmatrix} \begin{pmatrix} 4 & 1 \end{pmatrix}$$

$$= \begin{pmatrix} 2 & 4 \\ 4 & 8 \end{pmatrix} + \begin{pmatrix} 2 & 3 \\ 4 & 6 \end{pmatrix} + \begin{pmatrix} 20 & 5 \\ 12 & 3 \end{pmatrix}$$

(b) **Hint.** For a pair of vectors $\mathbf{x}$ and $\mathbf{y}$ the transpose of the outer product $\mathbf{xy}^T$ is

$$(\mathbf{xy}^T)^T = (\mathbf{y}^T)^T\mathbf{x}^T = \mathbf{yx}^T$$

8. **Hint.** AX and B are equal if and only if their corresponding column vectors are equal.

9. (b) **Hint.** $A\mathbf{e}_j = \mathbf{a}_j$ for $j = 1, \ldots, n$.

10. (b) **Hint.** Let $X = U\Sigma$. If $A = U\Sigma V^T$ then $A = XV^T$, so A can be expressed as an outer product expansion of XV^T. How are the column vectors of X related to the column vectors of U?

11. **Hint.** You need to determine a matrix C so that the block multiplication of

$$\begin{pmatrix} A_{11}^{-1} & C \\ O & A_{22}^{-1} \end{pmatrix} \begin{pmatrix} A_{11} & A_{12} \\ O & A_{22} \end{pmatrix} = \begin{pmatrix} I & A_{11}^{-1}A_{12} + CA_{22} \\ O & I \end{pmatrix}$$

will equal the $2n \times 2n$ identity matrix

$$I_{2n} = \begin{pmatrix} I & O \\ O & I \end{pmatrix}$$

Once you have found C check that the product of the matrices in the reverse order will also equal I_{2n}.

15. **Hint.** The block form of S^{-1} is

$$S^{-1} = \begin{pmatrix} I & -A \\ O & I \end{pmatrix}$$

16. **Hint.** Just plug in for B and C and multiply things out.

17. **Hint.** In order for the block multiplication to work we must have

$$XB = S \qquad \text{and} \qquad YM = T$$

Fortunately both B and M are nonsingular.

18. **Solution.**
(a)

$$BC = \begin{pmatrix} b_1 \\ b_2 \\ \vdots \\ b_n \end{pmatrix}(c) = \begin{pmatrix} b_1 c \\ b_2 c \\ \vdots \\ b_n c \end{pmatrix} = \begin{pmatrix} cb_1 \\ cb_2 \\ \vdots \\ cb_n \end{pmatrix} = c\mathbf{b}$$

(b)

$$A\mathbf{x} = (\mathbf{a}_1, \mathbf{a}_2, \ldots, \mathbf{a}_n) \begin{pmatrix} x_1 \\ x_2 \\ \vdots \\ x_n \end{pmatrix}$$

$$= \mathbf{a}_1(x_1) + \mathbf{a}_2(x_2) + \cdots + \mathbf{a}_n(x_n)$$

(c) It follows from parts (a) and (b) that

$$Ax = \mathbf{a}_1(x_1) + \mathbf{a}_2(x_2) + \cdots + \mathbf{a}_n(x_n)$$
$$= x_1\mathbf{a}_1 + x_2\mathbf{a}_2 + \cdots + x_n\mathbf{a}_n$$

19. The hint for this exercise is given in the textbook.

20. Hint. The hint for Exercise 19 also applies to this exercise.

MATLAB EXERCISES

1. In parts (a), (b), (c) it should turn out that $A1 = A4$ and $A2 = A3$. In part (d) $A1 = A3$ and $A2 = A4$. Exact equality will not occur in parts (c) and (d) because of roundoff error.

3. (a) **Hint.** See Theorem 1.4.2.

(b) **Hint.** How are the columns of B related to $\mathbf{x}$ and what does this imply about the columns of AB?

4. Hint. The key to answering the questions posed here is provided by Theorem 1.4.2.

8. Solution. The change in part (b) should not have a significant effect on the survival potential for the turtles. The change in part (c) will effect the $(2, 2)$ and $(3, 2)$ of the Leslie matrix. The new values will be $l_{22} = 0.9540$ and $l(3, 2) = 0.0101$. With these values the Leslie population model should predict that the survival period will double but the turtles will still eventually die out.

10. Solution.

(a)

$$A^{2k} = \begin{pmatrix} I & kB \\ kB & I \end{pmatrix}$$

This can be proved using mathematical induction. In the case $k = 1$

$$A^2 = \begin{pmatrix} O & I \\ I & B \end{pmatrix} \begin{pmatrix} O & I \\ I & B \end{pmatrix} = \begin{pmatrix} I & B \\ B & I \end{pmatrix}$$

If the result holds for $k = m$

$$A^{2m} = \begin{pmatrix} I & mB \\ mB & I \end{pmatrix}$$

then

$$A^{2m+2} = A^2 A^{2m}$$

$$= \begin{pmatrix} I & B \\ B & I \end{pmatrix} \begin{pmatrix} I & mB \\ mB & I \end{pmatrix}$$

$$= \begin{pmatrix} I & (m+1)B \\ (m+1)B & I \end{pmatrix}$$

It follows by mathematical induction that the result holds for all positive integers k.

(b)

$$A^{2k+1} = AA^{2k} = \begin{pmatrix} O & I \\ I & B \end{pmatrix} \begin{pmatrix} I & kB \\ kB & I \end{pmatrix} = \begin{pmatrix} kB & I \\ I & (k+1)B \end{pmatrix}$$

11. (b) **Hint.**

$$LDL^T = \begin{pmatrix} I & O \\ E & I \end{pmatrix} \begin{pmatrix} B_{11} & O \\ O & F \end{pmatrix} \begin{pmatrix} I & E^T \\ O & I \end{pmatrix}$$

$$= \begin{pmatrix} B_{11} & B_{11}E^T \\ EB_{11} & EB_{11}E^T + F \end{pmatrix}$$

Plug in

$$E = B_{21}B_{11}^{-1} \quad \text{and} \quad F = B_{22} - B_{21}B_{11}^{-1}B_{12}$$

and show that the three matrices multiply out to B.

CHAPTER TEST A

1. **Hint.** Can you make up an example of an inconsistent system that is in row echelon form and has free variables?

2. **Hint.** For a homogeneous system there is always one solution that is very obvious.

3. **Hint.** See Theorem 1.4.2 part (c).

4. **Hint.** Try making up some examples to see if this works.

5. **Hint.** See Theorem 1.3.3. By that theorem, if A and B are nonsingular matrices, then AB must also be nonsingular, however, $(AB)^{-1} = B^{-1}A^{-1}$. Why is it necessary to reverse the order of the two matrices to get the inverse formula to work? Make up an example of two nonsingular 2×2 matrices for which $(AB)^{-1} \neq A^{-1}B^{-1}$.

6. **Hint.** It should be easy to make up a counterexample for this statement. To see why the formula doesn't work multiply out the expression

$$(A - B)^2 = (A - B)(A - B)$$

and remember that matrix multiplication is not commutative.

7. **Hint.** If $AB = AC$ and A is nonsingular then you should be able to give a one line proof that B must equal C. What about the case where A is singular and $A \neq O$. In this case, can you find B and C such that $AB = AC$ and $B \neq C$. Try making up an example using 2×2 matrices.

8. **Hint.** Make up some elementary matrices and multiply them together. Do you always end up with an elementary matrix?

9. **Hint.** How are the rows of $A = \mathbf{x}\mathbf{y}^T$ related to $\mathbf{y}^T$ and what does this relation imply about any row echelon of the matrix?

10. **Hint.** Given the form of $\mathbf{b}$, you should be able to determine one solution. So the system must be consistent. How many solutions there are will depend upon how many ways you can write $\mathbf{b}$ as a linear combination of the column vectors of A.

CHAPTER TEST B

2. (c) **Hint.** You can make use of a theorem from Section 2 or you can give a geometric explanation.

3. (c) Show that if A is nonsingular the system must have exactly one solution.

4. **Hint.** Use the consistency theorem.

5. (a) **Hint.** What row operation is performed on A to transform it to B?
   (b) **Hint.** What column operation is performed on A to transform it to C?

6. **Hint.** Refer to a theorem.

7. **Hint.** Make use of Theorem 1.4.2.

9. **Hint.** Try some examples.

11. **Hint.** The block structure of A resembles the form of an elementary matrix of type III.

12. (a) **Hint.** The column partition of A must match up with the row partition of B.

Chapter 2
Determinants

1 THE DETERMINANT OF A MATRIX

Overview

With each square matrix it is possible to associate a real number called the determinant of the matrix. The value of this number will tell us whether or not the matrix is singular. In this section we give an inductive definition of the determinant of a matrix in terms of determinants of submatrices. In the case that A is a 2×2 or a 3×3 matrix we show that A is nonsingular if and only if its determinant is nonzero. The more general result for $n \times n$ matrices is proved in Section 2 of the textbook.

Important Concepts

1. **Minor and Cofactor.** Let $A = (a_{ij})$ be an $n \times n$ matrix and let M_{ij} denote the $(n-1) \times (n-1)$ matrix obtained from A by deleting the row and column containing a_{ij}. The determinant of M_{ij} is called the *minor* of a_{ij}. We define the *cofactor* A_{ij} of a_{ij} by

$$A_{ij} = (-1)^{i+j} \det(M_{ij})$$

2. **Definition of Determinant.** The *determinant* of an $n \times n$ matrix A, denoted $\det(A)$, is a scalar associated with the matrix A that is defined inductively as follows:

$$\det(A) = \begin{cases} a_{11} & \text{if } n = 1 \\ a_{11}A_{11} + a_{12}A_{12} + \cdots + a_{1n}A_{1n} & \text{if } n > 1 \end{cases}$$

where A_{1j} is the (1,j) cofactor of A for $j = 1, \ldots, n$.

3. **Cofactor Expansion.** Although the determinant is defined in terms of the cofactors associated with the first row, it can be expressed as a *cofactor expansion* using any row or column. The cofactor expansion of $\det(A)$ along the ith row is given by

$$\det(A) = a_{i1}A_{i1} + a_{i2}A_{i2} + \cdots + a_{in}A_{in}$$

and the expansion along the jth column is

$$\det(A) = a_{1j}A_{1j} + a_{2j}A_{2j} + \cdots + a_{nj}A_{nj}$$

27

Important Theorems

▶ THEOREM 2.1.1. *If A is an $n \times n$ matrix with $n \geq 2$, then $\det(A)$ can be expressed as a cofactor expansion using any row or column of A.*

$$\det(A) = a_{i1}A_{i1} + a_{i2}A_{i2} + \cdots + a_{in}A_{in}$$
$$= a_{1j}A_{1j} + a_{2j}A_{2j} + \cdots + a_{nj}A_{nj}$$

for $i = 1, \ldots, n$ and $j = 1, \ldots, n$.

▶ THEOREM 2.1.2. *If A is an $n \times n$ matrix, then $\det(A^T) = \det(A)$.*

▶ THEOREM 2.1.3. *If A is an $n \times n$ triangular matrix, the determinant of A equals the product of the diagonal elements of A.*

Exercises: Solutions and Hints

7. Solution. Given that $a_{11} = 0$ and $a_{21} \neq 0$, let us interchange the first two rows of A and also multiply the third row through by $-a_{21}$. We end up with the matrix

$$\begin{pmatrix} a_{21} & a_{22} & a_{23} \\ 0 & a_{12} & a_{13} \\ -a_{21}a_{31} & -a_{21}a_{32} & -a_{21}a_{33} \end{pmatrix}$$

Now if we add a_{31} times the first row to the third, we obtain the matrix

$$\begin{pmatrix} a_{21} & a_{22} & a_{23} \\ 0 & a_{12} & a_{13} \\ 0 & a_{31}a_{22} - a_{21}a_{32} & a_{31}a_{23} - a_{21}a_{33} \end{pmatrix}$$

This matrix will be row equivalent to I if and only if

$$\begin{vmatrix} a_{12} & a_{13} \\ a_{31}a_{22} - a_{21}a_{32} & a_{31}a_{23} - a_{21}a_{33} \end{vmatrix} \neq 0$$

Thus the original matrix A will be row equivalent to I if and only if

$$a_{12}a_{31}a_{23} - a_{12}a_{21}a_{33} - a_{13}a_{31}a_{22} + a_{13}a_{21}a_{32} \neq 0$$

8. Solution. Theorem 2.1.3 states that if A is an $n \times n$ triangular matrix, then the determinant of A equals the product of the diagonal elements of A. The proof is by induction on n. In the case $n = 1$, $A = (a_{11})$ and $\det(A) = a_{11}$. Assume the result holds for all $k \times k$ triangular matrices and let A be a $(k+1) \times (k+1)$ lower triangular matrix. (It suffices to prove the theorem for lower triangular matrices since $\det(A^T) = \det(A)$.) If $\det(A)$ is expanded by cofactors using the first row of A we get

$$\det(A) = a_{11} \det(M_{11})$$

where M_{11} is the $k \times k$ matrix obtained by deleting the first row and column of A. Since M_{11} is lower triangular we have

$$\det(M_{11}) = a_{22}a_{33} \cdots a_{k+1,k+1}$$

and consequently

$$\det(A) = a_{11}a_{22} \cdots a_{k+1,k+1}$$

10. **Solution.** In the case $n = 1$, if A is a matrix of the form

$$\begin{pmatrix} a & b \\ a & b \end{pmatrix}$$

then $\det(A) = ab - ab = 0$. Suppose that the result holds for $(k+1) \times (k+1)$ matrices and that A is a $(k+2) \times (k+2)$ matrix whose ith and jth rows are identical. Expand $\det(A)$ by factors along the mth row where $m \neq i$ and $m \neq j$.

$$\det(A) = a_{m1} \det(M_{m1}) + a_{m2} \det(M_{m2}) + \cdots + a_{m,k+2} \det(M_{m,k+2}).$$

Each M_{ms}, $1 \leq s \leq k + 2$, is a $(k + 1) \times (k + 1)$ matrix containing two identical rows. Thus by the induction hypothesis

$$\det(M_{ms}) = 0 \qquad (1 \leq s \leq k + 2)$$

and consequently $\det(A) = 0$.

11. (a) **Hint.** Consider some examples with 2×2 matrices.
    (b) **Hint.** Just use brute force. Compute $\det(A)$ and $\det(B)$ in terms of $a_{11}, a_{12}, a_{21}, a_{22}, b_{11}, b_{12}, b_{21}, b_{22}$ and compare the result to what you get if you first compute AB and then take its determinant.
    (c) **Hint.** Make use of the result from part (b).

---

2   PROPERTIES OF DETERMINANTS

Overview

This section deals with the effects of row operations on the value of the determinant of a matrix. For $n > 3$, determinants can be computed more efficiently using a method based on elimination. The idea of the method is to try to use row operations I and III to reduce a matrix A to upper triangular form and to keep track of how many times rows are interchanged in the process. If the last row zeros out completely, then A is singular and $\det(A) = 0$. If A is nonsingular, then the reduction process will produce an upper triangular matrix U with nonzero diagonal entries. If row operation I is used k times in the process, then

$$\det(A) = (-1)^k \det(U) = (-1)^k u_1 u_1 \cdots u_n$$

Important Concepts

1. Effect of Row Operation I.

If E is an elementary matrix of type I, then $\det(EA) = -\det(A)$.

2. Effect of Row Operation II.

If E is an elementary matrix of type II, formed by multiplying a row of I by a nonzero scalar c, then $\det(EA) = c\det(A)$.

3. Effect of Row Operation III.

If E is an elementary matrix of type III, then $\det(EA) = \det(A)$.

Important Theorems

▶ THEOREM 2.2.2. *An $n \times n$ matrix A is singular if and only if*

$$\det(A) = 0$$

▶ THEOREM 2.2.3. *If A and B are $n \times n$ matrices, then*

$$\det(AB) = \det(A)\det(B)$$

Exercises: Solutions and Hints

2. (a) **Solution.** We reduce A to upper triangular form using row operations I and III. Each time we interchange rows we change the sign of the determinant. (This only happens once at the very first step of the process.)

$$\det(A) = \begin{vmatrix} 0 & 1 & 2 & 3 \\ 1 & 1 & 1 & 1 \\ -2 & -2 & 3 & 3 \\ 1 & 2 & -2 & -3 \end{vmatrix}$$

$$= -\begin{vmatrix} 1 & 1 & 1 & 1 \\ 0 & 1 & 2 & 3 \\ -2 & -2 & 3 & 3 \\ 1 & 2 & -2 & -3 \end{vmatrix}$$

$$= -\begin{vmatrix} 1 & 1 & 1 & 1 \\ 0 & 1 & 2 & 3 \\ 0 & 0 & 5 & 5 \\ 0 & 1 & -3 & -4 \end{vmatrix}$$

$$= -\begin{vmatrix} 1 & 1 & 1 & 1 \\ 0 & 1 & 2 & 3 \\ 0 & 0 & 5 & 5 \\ 0 & 0 & -5 & -7 \end{vmatrix}$$

$$= -\begin{vmatrix} 1 & 1 & 1 & 1 \\ 0 & 1 & 2 & 3 \\ 0 & 0 & 5 & 5 \\ 0 & 0 & 0 & -2 \end{vmatrix}$$

$$= -1 \cdot 1 \cdot 5 \cdot (-2)$$

$$= 10$$

5. **Hint.** What is the effect on $\det(A)$ if you multiply just one row of A by α? When you do the scalar multiplication αA you are multiplying all of the rows of A by α.

6. **Hint.** Use Theorem 2.2.3 to prove the result.

9. (b) 18;   (d) -6;   (f) -3

10. **Hint.** Look at the list of important concepts for this section.

11. **Hint.** Show that if A is a $n \times n$ whose entries are real numbers then $\det(A^2) \geq 0$.

12. (a) **Hint.** Use elimination to zero out the $(2,1)$ and $(3,1)$ entries of V before computing its determinant.

14. **Hint.** Use Theorem 2.2.3.

15. If $AB = I$, use determinants to show that A must be nonsingular and then show that $B = A^{-1}$.

16. **Hint.** Make use of the result from Exercise 5.

17. **Hint.** Subtract $c = \det(A)/A_{nn}$ from the (n,n) entry of A and then do a cofactor expansion along the last row of the matrix.

20. **Solution.** In the elimination method the matrix is reduced to triangular form and the determinant of the triangular matrix is calculated by multiplying its diagonal elements. At the first step of the reduction process the first

row is multiplied by $m_{i1} = -a_{i1}/a_{11}$ and then added to the ith row. This requires 1 division, $n-1$ multiplications and $n-1$ additions. However, this row operation is carried out for $i = 2, \ldots, n$. Thus the first step of the reduction requires $n-1$ divisions, $(n-1)^2$ multiplications and $(n-1)^2$ additions. At the second step of the reduction this same process is carried out on the $(n-1) \times (n-1)$ matrix obtained by deleting the first row and first column of the matrix obtained from step 1. The second step of the elimination process requires $n-2$ divisions, $(n-2)^2$ multiplications, and $(n-2)^2$ additions. After $n-1$ steps the reduction to triangular form will be complete. It will require:

$$(n-1) + (n-2) + \cdots + 1 = \frac{n(n-1)}{2} \text{ divisions}$$

$$(n-1)^2 + (n-2)^2 + \cdots + 1^2 = \frac{n(2n-1)(n-1)}{6} \text{ multiplications}$$

$$(n-1)^2 + (n-2)^2 + \cdots + 1^2 = \frac{n(2n-1)(n-1)}{6} \text{ additions}$$

It takes $n-1$ additional multiplications to calculate the determinant of the triangular matrix. Thus the calculation $\det(A)$ by the elimination method requires:

$$\frac{n(n-1)}{2} + \frac{n(2n-1)(n-1)}{6} + (n-1) = \frac{(n-1)(n^2+n+3)}{3}$$

multiplications and divisions and $\dfrac{n(2n-1)(n-1)}{6}$ additions.

3 | CRAMER'S RULE

Overview

In this section the book develops a rule for expressing the solution to a linear system in terms of determinants. The rule is often used for 2×2 and 3×3 linear systems, but it is not that practical to use for larger systems.

Important Concepts

1. **Adjoint of a matrix.** The *adjoint* of an $n \times n$ matrix A is the matrix defined by

$$\text{adj } A = \begin{pmatrix} A_{11} & A_{21} & \cdots & A_{n1} \\ A_{12} & A_{22} & \cdots & A_{n2} \\ \vdots & & & \\ A_{1n} & A_{2n} & \cdots & A_{nn} \end{pmatrix}$$

where A_{ij} denotes the (i, j) cofactor of A. Thus to form the adjoint, we must replace each entry of A by its cofactor and then transpose the resulting matrix.

2. If A is a nonsingular matrix then $A^{-1} = \dfrac{1}{\det(A)} \operatorname{adj}(A)$.

Important Theorems

▶ THEOREM 2.3.1 **(Cramer's Rule)** *Let A be an $n \times n$ nonsingular matrix and let $\mathbf{b} \in R^n$. Let A_i be the matrix obtained by replacing the ith column of A by $\mathbf{b}$. If $\mathbf{x}$ is the unique solution to $A\mathbf{x} = \mathbf{b}$, then*

$$x_i = \frac{\det(A_i)}{\det(A)} \quad for \quad i = 1, 2, \ldots, n$$

Exercises: Solutions and Hints

1. (c) **Solution.** The cofactors of the entries of A are:

$$\begin{array}{lll} A_{11} = -3, & A_{12} = 0 & A_{13} = 6 \\ A_{21} = 5, & A_{22} = 1 & A_{23} = -8 \\ A_{31} = 2, & A_{12} = 1 & A_{13} = -5 \end{array}$$

It follows that

$$\det A = a_{11}A_{11} + a_{12}A_{12} = a_{13}A_{13} = 1 \cdot (-3) + 3 \cdot 0 + 1 \cdot 6 = 3$$

$$\operatorname{adj} A = \begin{pmatrix} -3 & 0 & 6 \\ 5 & 1 & -8 \\ 2 & 1 & -5 \end{pmatrix}^T = \begin{pmatrix} -3 & 5 & 2 \\ 0 & 1 & 1 \\ 6 & -8 & -5 \end{pmatrix}$$

and

$$A^{-1} = \frac{1}{\det(A)} \operatorname{adj} A = \frac{1}{3} \operatorname{adj} A = \begin{pmatrix} -1 & \frac{5}{3} & \frac{2}{3} \\ 0 & \frac{1}{3} & \frac{1}{3} \\ 2 & -\frac{8}{3} & -\frac{5}{3} \end{pmatrix}$$

2. (c) **Solution.** The determinant of coefficient matrix A is

$$\det(A) = \begin{vmatrix} 2 & 1 & -3 \\ 4 & 5 & 1 \\ -2 & -1 & 4 \end{vmatrix} = 6$$

so A is nonsingular and the system has a unique solution. For $j = 1, 2, 3$, let A_j denote the matrix obtained by replacing the jth column of A by the right hand side **b**.

$$\det(A_1) = \begin{vmatrix} 0 & 1 & -3 \\ 8 & 5 & 1 \\ 2 & -1 & 4 \end{vmatrix} = 24$$

$$\det(A_2) = \begin{vmatrix} 2 & 0 & -3 \\ 4 & 8 & 1 \\ -2 & 2 & 4 \end{vmatrix} = -12$$

$$\det(A_3) = \begin{vmatrix} 2 & 1 & 0 \\ 4 & 5 & 8 \\ -2 & -1 & 2 \end{vmatrix} = 12$$

By Cramer's rule

$$x_1 = \frac{\det A_1}{\det A} = \frac{24}{6} = 4$$
$$x_2 = \frac{\det A_2}{\det A} = -\frac{12}{6} = -2$$
$$x_3 = \frac{\det A_3}{\det A} = \frac{12}{6} = 2$$

The solution to the system is $\mathbf{x} = (4, -2, 2)^T$.

7. Hint. The solution of the linear system $I\mathbf{x} = \mathbf{b}$ is $\mathbf{x} = \mathbf{b}$.

11. Solution. If $A = O$, then $\operatorname{adj} A$ is also the zero matrix and hence is singular. If A is singular and $A \neq O$, then

$$A \operatorname{adj} A = \det(A)I = 0I = O$$

If $\mathbf{a}^T$ is any nonzero row vector of A, then

$$\mathbf{a}^T \operatorname{adj} A = \mathbf{0}^T \qquad \text{or} \qquad (\operatorname{adj} A)^T \mathbf{a} = \mathbf{0}$$

It follows from Theorem 1.4.3 that $(\operatorname{adj} A)^T$ is singular. Since

$$\det(\operatorname{adj} A) = \det[(\operatorname{adj} A)^T] = 0$$

it follows that $\operatorname{adj} A$ is singular.

12. Hint. If $\det(A) = 1$ show that

$$\operatorname{adj} A = A^{-1}$$

and then apply the result from Exercise 10.

13. Hint. The (j, i) entry of Q^T is q_{ij}. Determine an expression for the (j, i) entry of Q^{-1} involving a cofactor of Q.

MATLAB EXERCISES

5. The matrix U is very sensitive to roundoff error, i.e., small roundoff errors may cause drastic changes in computed solutions. When matrices are so sensitive to roundoff error that computed solutions may contain no digits of accuracy we say that the matrix is singular with respect to the machine precision used by MATLAB. For such matrices the computed values of the determinants will not be accurate. Both U^T and UU^T fall into this category and consequently we cannot expect to get even a single digit of accuracy in the computed values of $\det(U^T)$ and $\det(UU^T)$. On the other hand, since U is upper triangular, the computed value of $\det(U)$ is the product of its diagonal entries. This value should be accurate to the machine precision.

6. Since $A\mathbf{x} = \mathbf{0}$ and $\mathbf{x} \neq \mathbf{0}$, the matrix must be singular. However, there may be no indication of this if the computations are done in floating point arithmetic. To compute the determinant MATLAB does Gaussian elimination to reduce the matrix to upper triangular form U and then multiplies the diagonal entries of U. In this case the product $u_{11}u_{22}u_{33}u_{44}u_{55}$ has magnitude on the order of 10^{14}. If the computed value of u_{66} has magnitude of the order 10^{-k} and $k \leq 14$, then MATLAB will round the result to a nonzero integer. (MATLAB knows that if you started with an integer matrix, you should end up with an integer value for the determinant.) In general if the determinant is computed in floating point arithmetic, then you cannot expect it to be a reliable indicator of whether or not a matrix is nonsingular.

CHAPTER TEST A

1. Hint. See Theorem 2.2.3.

2. Hint. Try some examples with 2×2 matrices.

3. Hint. See Exercise 5 in Section 2.

4. Hint. See Theorems 2.1.2 and 2.2.3.

5. Hint. An easy way to make up a counterexample would be to use diagonal matrices for A and B.

6. Hint. Make use of Theorem 2.2.3 to prove the result.

7. Hint. See Theorems 2.1.3 and 2.2.2.

8. Hint. See Theorem 1.4.2.

9. Hint. What are the effects of row operations I and II on the value of the determinant?

10. Hint. How is $\det(A^k)$ related to $\det(A)$?

CHAPTER TEST B

1. (a) $\det(\frac{1}{2}A) = \det(\frac{1}{2}IA) = \det(\frac{1}{2}I)\det(A) = \frac{1}{8} \cdot 4 = \frac{1}{2}$

2. (a) $\det(A) = x^3 - x$

4. Hint. Use Theorem 2.1.2

5. Hint. How are the determinants of S and S^{-1} related?

7. **Hint.** Use Theorem 1.4.2.

8. **Hint.** Show that $\det(A) = x_1 x_2 \det(B)$ where B is a matrix whose first two rows are equal.

9. **Hint.** Use Theorem 1.4.2.

10. **Hint.** If A has integer entries then $\text{adj}\, A$ will also have integer entries.

Chapter 3

Vector Spaces

1 | DEFINITION AND EXAMPLES

Overview

This section of the book presents the definition of a type of mathematical system called a vector space and provides examples of some of the most important vector spaces. The primary vector spaces that get most of our attention in this course are the Euclidean vector spaces R^n. The elements of R^n are column vectors, i.e., $n \times 1$ matrices. More generally you could talk about vector spaces $R^{m \times n}$ whose elements or vectors are $m \times n$ matrices. Other important vector spaces are those whose vectors consist of a set of functions.

Important Concepts

1. **Vector Space.** Let V be a set on which the operations of addition and scalar multiplication are defined. By this we mean that with each pair of elements $\mathbf{x}$ and $\mathbf{y}$ in V, one can associate a unique element $\mathbf{x}+\mathbf{y}$ that is also in V, and with each element $\mathbf{x}$ in V and each scalar α, one can associate a unique element $\alpha\mathbf{x}$ in V. The set V together with the operations of addition and scalar multiplication is said to form a *vector space* if the following axioms are satisfied.

    A1. $\mathbf{x}+\mathbf{y} = \mathbf{y}+\mathbf{x}$ for any $\mathbf{x}$ and $\mathbf{y}$ in V.

    A2. $(\mathbf{x}+\mathbf{y})+\mathbf{z} = \mathbf{x}+(\mathbf{y}+\mathbf{z})$ for any $\mathbf{x},\mathbf{y},\mathbf{z}$ in V.

    A3. There exists an element $\mathbf{0}$ in V such that $\mathbf{x}+\mathbf{0} = \mathbf{x}$ for each $\mathbf{x} \in V$.

    A4. For each $\mathbf{x} \in V$, there exists an element $-\mathbf{x}$ in V such that $\mathbf{x}+(-\mathbf{x}) = \mathbf{0}$.

    A5. $\alpha(\mathbf{x}+\mathbf{y}) = \alpha\mathbf{x}+\alpha\mathbf{y}$ for each scalar α and any $\mathbf{x}$ and $\mathbf{y}$ in V.

    A6. $(\alpha+\beta)\mathbf{x} = \alpha\mathbf{x}+\beta\mathbf{x}$ for any scalars α and β and any $\mathbf{x} \in V$.

    A7. $(\alpha\beta)\mathbf{x} = \alpha(\beta\mathbf{x})$ for any scalars α and β and any $\mathbf{x} \in V$.

    A8. $1 \cdot \mathbf{x} = \mathbf{x}$ for all $\mathbf{x} \in V$.

    The elements of V are called *vectors* and are usually denoted by letters from the end of the alphabet: $\mathbf{u}$, $\mathbf{v}$, $\mathbf{w}$, $\mathbf{x}$, $\mathbf{y}$, and $\mathbf{z}$. The term *scalar* will generally refer to a real number, although in some cases it will be used to refer to complex numbers.

    It is probably not necessary to memorize this entire definition, however, what you should keep in mind is that a vector space has three components: (i) a set of elements called vectors, (ii) two operations, scalar multiplication (i.e., multiplication of a scalar times a vector) and vector addition (addition of two vectors), (iii) a set of axioms that the mathematical system must satisfy.

2. **Closure properties.** An important component of the vector space definition is the *closure properties* of the two operations. These properties can be summarized as follows:

    C1. If $\mathbf{x} \in V$ and α is a scalar, then $\alpha\mathbf{x} \in V$.

    C2. If $\mathbf{x},\mathbf{y} \in V$, then $\mathbf{x}+\mathbf{y} \in V$.

3. **Euclidean vector spaces.** Perhaps the most important vector spaces are the Euclidean vector spaces R^n. The elements of R^n are column vectors with n entries that all must be real numbers. Scalar multiplication and addition are defined in the standard way. (See Chapter 1, Section 3.)

4. $R^{m \times n}$. The elements in the vector space $R^{m \times n}$ are all $m \times n$ matrices with real entries. Scalar multiplication and addition are defined in the standard way.

5. $C[a, b]$. The elements in the vector space $C[a, b]$ are all real valued functions that are continuous in the closed interval $[a, b]$. The sum of two vectors f and g is defined to be the function $f + g$ where

$$(f + g)(x) = f(x) + g(x)$$

for all x in $[a, b]$. Scalar multiplication of a vector f by a scalar α is defined to be the function αf where

$$(\alpha f)(x) = \alpha f(x)$$

for all x in $[a, b]$.

6. P_n. The elements in the vector space P_n are all polynomials of degree less than n. Addition and scalar multiplication for the vector space P_n are defined in the same manner as they are for $C[a, b]$

Important Theorem

▶ THEOREM 3.1.1. *If V is a vector space and $\mathbf{x}$ is any element of V, then*

(i) $0\mathbf{x} = \mathbf{0}$.

(ii) $\mathbf{x} + \mathbf{y} = \mathbf{0}$ *implies that* $\mathbf{y} = -\mathbf{x}$ *(i.e., the additive inverse of $\mathbf{x}$ is unique).*

(iii) $(-1)\mathbf{x} = -\mathbf{x}$.

Exercises: Solutions and Hints

3. **Solution.** To show that C is a vector space we must show that all eight axioms are satisfied.

A1. $(a + bi) + (c + di) = (a + c) + (b + d)i$
$$= (c + a) + (d + b)i$$
$$= (c + di) + (a + bi)$$

A2. $(\mathbf{x} + \mathbf{y}) + \mathbf{z} = [(x_1 + x_2 i) + (y_1 + y_2 i)] + (z_1 + z_2 i)$
$$= (x_1 + y_1 + z_1) + (x_2 + y_2 + z_2)i$$
$$= (x_1 + x_2 i) + [(y_1 + y_2 i) + (z_1 + z_2 i)]$$
$$= \mathbf{x} + (\mathbf{y} + \mathbf{z})$$

A3. $(a + bi) + (0 + 0i) = (a + bi)$

A4. If $\mathbf{z} = a + bi$ then define $-\mathbf{z} = -a - bi$. It follows that
$$\mathbf{z} + (-\mathbf{z}) = (a + bi) + (-a - bi) = 0 + 0i = \mathbf{0}$$

A5. $\alpha[(a + bi) + (c + di)] = (\alpha a + \alpha c) + (\alpha b + \alpha d)i$
$$= \alpha(a + bi) + \alpha(c + di)$$

A6. $(\alpha + \beta)(a + bi) = (\alpha + \beta)a + (\alpha + \beta)bi$
$$= \alpha(a + bi) + \beta(a + bi)$$

A7. $(\alpha\beta)(a + bi) = (\alpha\beta)a + (\alpha\beta)bi$
$$= \alpha(\beta a + \beta bi)$$

A8. $1 \cdot (a + bi) = 1 \cdot a + 1 \cdot bi = a + bi$

4. **Hint.** As in the previous exercise you must check to see if the eight axioms hold. Theorem 1.3.1 guarantees that 5 of the axioms hold, so you need only show that the remaining 3 also hold.

5. **Solution.** Let f, g and h be arbitrary elements of $C[a, b]$.
A1. For all x in $[a, b]$

$$(f + g)(x) = f(x) + g(x) = g(x) + f(x) = (g + f)(x).$$

Therefore

$$f + g = g + f$$

A2. For all x in $[a, b]$,

$$
\begin{aligned}
[(f + g) + h](x) &= (f + g)(x) + h(x) \\
&= f(x) + g(x) + h(x) \\
&= f(x) + (g + h)(x) \\
&= [f + (g + h)](x)
\end{aligned}
$$

Therefore

$$(f + g) + h = f + (g + h)$$

A3. If $z(x)$ is identically 0 on $[a, b]$, then for all x in $[a, b]$

$$(f + z)(x) = f(x) + z(x) = f(x) + 0 = f(x)$$

Thus

$$f + z = f$$

A4. Define $-f$ by

$$(-f)(x) = -f(x) \quad \text{for all } x \text{ in } [a, b]$$

Since

$$(f + (-f))(x) = f(x) - f(x) = 0$$

for all x in $[a, b]$ it follows that

$$f + (-f) = z$$

A5. For each x in $[a, b]$

$$
\begin{aligned}
[\alpha(f + g)](x) &= \alpha f(x) + \alpha g(x) \\
&= (\alpha f)(x) + (\alpha g)(x)
\end{aligned}
$$

Thus

$$\alpha(f + g) = \alpha f + \alpha g$$

A6. For each x in $[a, b]$

$$
\begin{aligned}
[(\alpha + \beta)f](x) &= (\alpha + \beta)f(x) \\
&= \alpha f(x) + \beta f(x) \\
&= (\alpha f)(x) + (\beta f)(x)
\end{aligned}
$$

Therefore

$$(\alpha + \beta)f = \alpha f + \beta f$$

A7. For each x in $[a, b]$,

$$[(\alpha\beta)f](x) = \alpha\beta f(x) = \alpha[\beta f(x)] = [\alpha(\beta f)](x)$$

Therefore

$$(\alpha\beta)f = \alpha(\beta f)$$

A8. For each x in $[a, b]$

$$(1f)(x) = 1f(x) = f(x)$$

Therefore

$$1f = f$$

9. (a) **Hint.** Show first that if $\mathbf{y} = \beta\mathbf{0}$, then $\mathbf{y} + \mathbf{y} = \mathbf{y}$.
   (b) **Hint.** Show that if $\alpha\mathbf{x} = \mathbf{0}$ and $\alpha \neq 0$, then $\mathbf{x} = \mathbf{0}$.

10. **Hint.** Since addition is defined in the standard way, check out the axioms that involve scalar multiplication.

13. **Hint.** Four of the axioms fail to hold. To show it's not a vector space all you need to show is a single example where one of the axioms fails to hold.

14. **Hint.** Two of the axioms fail to hold.

15. **Solution.** If $\{a_n\}$, $\{b_n\}$, $\{c_n\}$ are arbitrary elements of S, then for each n

$$a_n + b_n = b_n + a_n$$

and

$$a_n + (b_n + c_n) = (a_n + b_n) + c_n$$

Hence

$$\{a_n\} + \{b_n\} = \{b_n\} + \{a_n\}$$
$$\{a_n\} + (\{b_n\} + \{c_n\}) = (\{a_n\} + \{b_n\}) + \{c_n\}$$

so Axioms 1 and 2 hold.

The zero vector is just the sequence $\{0, 0, \ldots\}$ and the additive inverse of $\{a_n\}$ is the sequence $\{-a_n\}$. The last four axioms all hold since

$$\alpha(a_n + b_n) = \alpha a_n + \alpha b_n$$
$$(\alpha + \beta)a_n = \alpha a_n + \beta a_n$$
$$\alpha\beta a_n = \alpha(\beta a_n)$$
$$1a_n = a_n$$

for each n. Thus all eight axioms hold and hence S is a vector space.

2 | SUBSPACES

Overview

If the vectors in a vector space W are all elements of a larger vector space V, then we say that W is a *subspace* of V. However, not every subset of V is a subspace. In order for a nonempty subset S to form of a subspace of V, the set must be closed under the operations of scalar multiplication and vector addition. One important example of a subspace is the set of all solutions to a homogeneous linear system $A\mathbf{x} = \mathbf{0}$. This subspace is referred to as the *nullspace* of A. If A is an $m \times n$ matrix, then its nullspace is a subspace of R^n. The section also introduces the important concepts of *span* and *spanning set*.

Important Concepts

1. **Subspace.** If S is a nonempty subset of a vector space V, and S satisfies the following conditions:

   (i) $\alpha \mathbf{x} \in S$ whenever $\mathbf{x} \in S$ for any scalar α

   (ii) $\mathbf{x} + \mathbf{y} \in S$ whenever $\mathbf{x} \in S$ and $\mathbf{y} \in S$

   then S is said to be a *subspace* of V.

2. **Nullspace of a matrix.** If A is an $m \times n$ matrix, then the set of all solutions to $A\mathbf{x} = \mathbf{0}$ is a subspace of R^n. This subspace is referred to the *nullspace* of A and is denoted $N(A)$.

3. **Linear combination of vectors.** Let $\mathbf{v}_1, \mathbf{v}_2, \ldots, \mathbf{v}_n$ be vectors in a vector space V. A sum of the form $\alpha_1 \mathbf{v}_1 + \alpha_2 \mathbf{v}_2 + \cdots + \alpha_n \mathbf{v}_n$, where $\alpha_1, \ldots, \alpha_n$ are scalars, is called a *linear combination* of $\mathbf{v}_1, \mathbf{v}_2, \ldots, \mathbf{v}_n$.

4. **Span of a set of vectors.** Let $\mathbf{v}_1, \mathbf{v}_2, \ldots, \mathbf{v}_k$ be vectors in a vector space V. The set of all linear combinations of $\mathbf{v}_1, \mathbf{v}_2, \ldots, \mathbf{v}_k$ is called the *span* of $\mathbf{v}_1, \ldots, \mathbf{v}_k$. The span of $\mathbf{v}_1, \ldots, \mathbf{v}_k$ is a subspace of V and it is denoted by $\mathrm{Span}(\mathbf{v}_1, \ldots, \mathbf{v}_k)$.

5. **Spanning set for a vector space.** The set $\{\mathbf{v}_1, \ldots, \mathbf{v}_n\}$ is a *spanning set* for V if and only if every vector in V can be written as a linear combination of $\mathbf{v}_1, \mathbf{v}_2, \ldots, \mathbf{v}_n$. In this case we say that the vectors $\mathbf{v}_1, \mathbf{v}_2, \ldots, \mathbf{v}_n$ *span* V.

Important Theorems

▶ THEOREM 3.2.1. *If* $\mathbf{v}_1, \mathbf{v}_2, \ldots, \mathbf{v}_n$ *are vectors in a vector space* V *and*

$$S = \mathrm{Span}(\mathbf{v}_1, \mathbf{v}_2, \ldots, \mathbf{v}_n)$$

then S *is a subspace of* V.

Exercises: Solutions and Hints

1. **Hint.** Be careful when you check if the operation is closed under vector addition.

4. (d) **Solution.** To find the nullspace we must solve the system $A\mathbf{x} = \mathbf{0}$. A homogeneous system with more unknowns than equations will have infinitely many solutions. So to solve the system we use row operations to transform $(A \,|\, \mathbf{0})$ to reduced row echelon form.

$$\left(\begin{array}{cccc|c} 1 & 1 & -1 & 2 & 0 \\ 2 & 2 & -3 & 1 & 0 \\ -1 & -1 & 0 & -5 & 0 \end{array} \right) \rightarrow \left(\begin{array}{cccc|c} 1 & 1 & -1 & 2 & 0 \\ 0 & 0 & -1 & -3 & 0 \\ 0 & 0 & -1 & -3 & 0 \end{array} \right)$$

$$\rightarrow \begin{pmatrix} 1 & 1 & -1 & 2 & | & 0 \\ 0 & 0 & -1 & -3 & | & 0 \\ 0 & 0 & 0 & 0 & | & 0 \end{pmatrix}$$

$$\rightarrow \begin{pmatrix} 1 & 1 & -1 & 2 & | & 0 \\ 0 & 0 & 1 & 3 & | & 0 \\ 0 & 0 & 0 & 0 & | & 0 \end{pmatrix}$$

$$\rightarrow \begin{pmatrix} 1 & 1 & 0 & 5 & | & 0 \\ 0 & 0 & 1 & 3 & | & 0 \\ 0 & 0 & 0 & 0 & | & 0 \end{pmatrix}$$

The free variables are x_2 and x_4. If we set $x_2 = a$ and $x_4 = b$, then $x_1 = -a - 5b$ and $x_3 = -3b$. The solution consists of all vectors of the form

$$\mathbf{x} = \begin{pmatrix} -a - 5b \\ a \\ -3b \\ b \end{pmatrix} = a \begin{pmatrix} -1 \\ 1 \\ 0 \\ 0 \end{pmatrix} + b \begin{pmatrix} -5 \\ 0 \\ -3 \\ 1 \end{pmatrix}$$

Thus the nullspace is the span of the vectors

$$\mathbf{x}_1 = \begin{pmatrix} -1 \\ 1 \\ 0 \\ 0 \end{pmatrix}, \qquad \mathbf{x}_2 = \begin{pmatrix} -5 \\ 0 \\ -3 \\ 1 \end{pmatrix}$$

8. (a) **Solution.**

If $B \in S_1$, then $AB = BA$. It follows that

$$A(\alpha B) = \alpha AB = \alpha BA = (\alpha B)A$$

and hence $\alpha B \in S_1$.

If B and C are in S_1, then

$$AB = BA \qquad \text{and} \qquad AC = CA$$

thus

$$A(B+C) = AB + AC = BA + CA = (B+C)A$$

and hence $B + C \in S_1$. Therefore S_1 is a subspace of $R^{2 \times 2}$.

(b) **Hint.** If $AB \neq BA$, then is $A(cB) \neq (cB)A$ for *all* scalars c?

10 (c) **Solution.** To determine whether the vectors span R^3 we need to see if any vector $(a, b, c)^T$ can be written as linear combination the vectors, i.e., we need to determine if it is possible to find scalars c_1, c_2, c_3, such that

$$c_1 \begin{pmatrix} 2 \\ 1 \\ -2 \end{pmatrix} + c_2 \begin{pmatrix} 3 \\ 2 \\ -2 \end{pmatrix} + c_3 \begin{pmatrix} 2 \\ 2 \\ 0 \end{pmatrix} = \begin{pmatrix} a \\ b \\ c \end{pmatrix}$$

This equation can be rewritten as a linear system.

$$\begin{aligned} 2c_1 + 3c_2 + 2c_3 &= a \\ c_1 + 2c_2 + 2c_3 &= b \\ -2c_1 - 2c_2 &= c \end{aligned}$$

The vectors will span if and only if this system of equations is consistent for every choice of the scalars a, b, c. To determine whether the system is consistent we use Gaussian elimination.

$$\begin{pmatrix} 2 & 3 & 2 & a \\ 1 & 2 & 2 & b \\ 2 & 2 & 0 & c \end{pmatrix} \rightarrow \begin{pmatrix} 1 & 2 & 2 & b \\ 2 & 3 & 2 & a \\ 2 & 2 & 0 & c \end{pmatrix}$$

$$\rightarrow \begin{pmatrix} 1 & 2 & 2 & b \\ 0 & -1 & -2 & a - 2b \\ 0 & -2 & -4 & c - 2b \end{pmatrix}$$

$$\rightarrow \begin{pmatrix} 1 & 2 & 2 & b \\ 0 & 1 & 2 & 2b - a \\ 0 & -2 & -4 & c - 2b \end{pmatrix}$$

$$\rightarrow \begin{pmatrix} 1 & 2 & 2 & b \\ 0 & 1 & 2 & 2b - a \\ 0 & 0 & 0 & c + 2b + 2a \end{pmatrix}$$

The system will be inconsistent for any choice of a, b, c, such that
$$c + 2b + 2a \neq 0$$
Therefore the vectors do not span R^3.

11 (a) **Hint.** $\mathbf{x} \in \mathrm{Span}(\mathbf{x}_1, \mathbf{x}_2)$ if and only if there exist scalars c_1 and c_2 such that
$$c_1 \mathbf{x}_1 + c_2 \mathbf{x}_2 = \mathbf{x}$$
Thus $\mathbf{x} \in \mathrm{Span}(\mathbf{x}_1, \mathbf{x}_2)$ if and only if the linear system $X\mathbf{c} = \mathbf{x}$ is consistent.

12. (b) **Solution.** If $\mathbf{x}_k \notin \mathrm{Span}(\mathbf{x}_1, \mathbf{x}_2, \ldots, \mathbf{x}_{k-1})$, then $\{\mathbf{x}_1, \mathbf{x}_2, \ldots, \mathbf{x}_{k-1}\}$ cannot be a spanning set. On the other hand if $\mathbf{x}_k \in \mathrm{Span}(\mathbf{x}_1, \mathbf{x}_2, \ldots, \mathbf{x}_{k-1})$, then
$$\mathrm{Span}(\mathbf{x}_1, \mathbf{x}_2, \ldots, \mathbf{x}_k) = \mathrm{Span}(\mathbf{x}_1, \mathbf{x}_2, \ldots, \mathbf{x}_{k-1})$$
and hence the $k - 1$ vectors will span the entire vector space.

13. **Hint.** Show that any matrix
$$A = \begin{pmatrix} a_{11} & a_{12} \\ a_{21} & a_{22} \end{pmatrix}$$
can be expressed as a linear combination of E_{11}, E_{12}, E_{21}, E_{22}.

16. **Hint.** If $S \neq \{0\}$ is a subspace of R^1, then it is closed under scalar multiplication.

17. **Hint.** You need to prove (a) implies (b), (b) implies (c), and (c) implies (a). All three implications should be simple to show.

19. **Hint.** Show that $S \cup T$ is not closed under vector addition.

20. **Solution.** If $\mathbf{z} \in U + V$, then $\mathbf{z} = \mathbf{u} + \mathbf{v}$ where $\mathbf{u} \in U$ and $\mathbf{v} \in V$. Since U and V are subspaces it follows that
$$\alpha \mathbf{u} \in U \quad \text{and} \quad \alpha \mathbf{v} \in V$$
for all scalars α. Thus
$$\alpha \mathbf{z} = \alpha \mathbf{u} + \alpha \mathbf{v}$$
is an element of $U + V$. If $\mathbf{z}_1$ and $\mathbf{z}_2$ are elements of $U + V$, then
$$\mathbf{z}_1 = \mathbf{u}_1 + \mathbf{v}_1 \quad \text{and} \quad \mathbf{z}_2 = \mathbf{u}_2 + \mathbf{v}_2$$
where $\mathbf{u}_1, \mathbf{u}_2 \in U$ and $\mathbf{v}_1, \mathbf{v}_2 \in V$. Since U and V are subspaces it follows that
$$\mathbf{u}_1 + \mathbf{u}_2 \in U \quad \text{and} \quad \mathbf{v}_1 + \mathbf{v}_2 \in V$$
Thus
$$\mathbf{z}_1 + \mathbf{z}_2 = (\mathbf{u}_1 + \mathbf{v}_1) + (\mathbf{u}_2 + \mathbf{v}_2) = (\mathbf{u}_1 + \mathbf{u}_2) + (\mathbf{v}_1 + \mathbf{v}_2)$$
is an element of $U + V$. Therefore $U + V$ is a subspace of W.

21. **Hint.** To show that two subspaces V and W are equal you must show that if $\mathbf{v}$ is any vector in V then $\mathbf{v} \in W$ and also that if $\mathbf{w}$ is any vector in W then $\mathbf{w} \in V$. Before trying to prove the statements look at some examples to see whether or not the statements appear to be true. Try taking V to be R^2 and take U, S and T to be subspaces that are each spanned by a single nonzero vector.

3 | LINEAR INDEPENDENCE

Overview

This section covers one of the most important topics in the course and in mathematics in general. Given a set of vectors, we wish to characterize dependency relations among the vectors. This is particularly important when solving homogeneous linear equations. For example, if the row echelon form of a matrix A has exactly three free variables then the linear system $A\mathbf{x} = \mathbf{0}$ will have infinite many solutions. The solution set can be characterized in terms of the span of three vectors. To do this we must find three solutions that are *linearly independent* in the sense that no one of the three is a linearly combination of the other two.

Important Concepts

1. **Linearly independent vectors.** The vectors $\mathbf{v}_1, \mathbf{v}_2, \ldots, \mathbf{v}_n$ in a vector space V are said to be *linearly independent* if

$$c_1\mathbf{v}_1 + c_2\mathbf{v}_2 + \cdots + c_n\mathbf{v}_n = \mathbf{0}$$

implies that all of the scalars $c_1, \ldots, c_n$ must all equal 0.

2. **Linearly dependent vectors.** The vectors $\mathbf{v}_1, \mathbf{v}_2, \ldots, \mathbf{v}_n$ in a vector space V are said to be *linearly dependent* if there exist scalars $c_1, c_2, \ldots, c_n$ not all zero such that

$$c_1\mathbf{v}_1 + c_2\mathbf{v}_2 + \cdots + c_n\mathbf{v}_n = \mathbf{0}$$

3. **Wronskian** Let $f_1, f_2, \ldots, f_n$ be functions in $C^{(n-1)}[a, b]$ and define the function $W[f_1, f_2, \ldots, f_n](x)$ on $[a, b]$ by

$$W[f_1, f_2, \ldots, f_n](x) = \begin{vmatrix} f_1(x) & f_2(x) & \cdots & f_n(x) \\ f_1'(x) & f_2'(x) & \cdots & f_n'(x) \\ \vdots & & & \\ f_1^{(n-1)}(x) & f_2^{(n-1)}(x) & \cdots & f_n^{(n-1)}(x) \end{vmatrix}$$

The function $W[f_1, f_2, \ldots, f_n]$ is called the *Wronskian* of $f_1, f_2, \ldots, f_n$.

Important Theorems

▶ THEOREM 3.3.1. *Let* $\mathbf{x}_1, \mathbf{x}_2, \ldots, \mathbf{x}_n$ *be n vectors in* R^n *and let*

$$\mathbf{x}_i = (x_{1i}, x_{2i}, \ldots, x_{ni})^T$$

for $i = 1, \ldots, n$. *If* $X = (\mathbf{x}_1, \mathbf{x}_2, \ldots, \mathbf{x}_n)$, *then the vectors* $\mathbf{x}_1, \mathbf{x}_2, \ldots, \mathbf{x}_n$ *will be linearly dependent if and only if* X *is singular.*

▶ THEOREM 3.3.2. *Let* $\mathbf{v}_1, \ldots, \mathbf{v}_n$ *be vectors in a vector space* V. *A vector* $\mathbf{v}$ *in* $\mathrm{Span}(\mathbf{v}_1, \ldots, \mathbf{v}_n)$ *can be written uniquely as a linear combination of* $\mathbf{v}_1, \ldots, \mathbf{v}_n$ *if and only if* $\mathbf{v}_1, \ldots, \mathbf{v}_n$ *are linearly independent.*

Exercises: Solutions and Hints

2. (e) **Solution.** To test for independence we need to determine if there are nonzero choices of c_1, c_2, such that

$$c_1 \begin{pmatrix} 1 \\ 1 \\ 3 \end{pmatrix} + c_2 \begin{pmatrix} 0 \\ 2 \\ 1 \end{pmatrix} = \begin{pmatrix} 0 \\ 0 \\ 0 \end{pmatrix}$$

To do this we rewrite the matrix equation as a linear system and transform the augmented matrix to row echelon form.

$$\left(\begin{array}{cc|c} 1 & 1 & 3 \\ 0 & 1 & 2 \\ 0 & 0 & 0 \end{array} \right) \rightarrow \left(\begin{array}{cc|c} 1 & 0 & 0 \\ 0 & 1 & 0 \\ 0 & 0 & 0 \end{array} \right)$$

Solving we see that $c_1 = c_2 = 0$ and hence the two vectors are linear independent. Actually it is fairly obvious that these vectors are linearly independent since neither is a multiple of the other (see Exercise 12).

4. (c) **Solution.** Let

$$A_1 = \begin{pmatrix} 1 & 0 \\ 0 & 1 \end{pmatrix}, \quad A_2 = \begin{pmatrix} 3 & 1 \\ 0 & 0 \end{pmatrix}, \quad A_3 = \begin{pmatrix} 2 & 3 \\ 0 & 2 \end{pmatrix}$$

It is easy to see that

$$2A_1 + 3A_2 - A_3 = O$$

Since the scalars $c_1 = 2$, $c_2 = 3$, and $c_3 = -1$ are not all 0, it follows that A_1, A_2, A_3 are linearly dependent.

5. (a) **Hint.** Show that $\mathbf{x}_{k+1} \in \mathrm{Span}(\mathbf{x}_1, \mathbf{x}_2, \ldots, \mathbf{x}_k)$ implies a dependency relation. On the other hand, if $\mathbf{x}_{k+1} \notin \mathrm{Span}(\mathbf{x}_1, \mathbf{x}_2, \ldots, \mathbf{x}_k)$ show that the $k + 1$ vectors will be linearly independent.

(b) **Hint.** Show that if $\mathbf{x}_1, \mathbf{x}_2, \ldots, \mathbf{x}_{k-1}$ were linearly dependent then this would imply that $\mathbf{x}_1, \mathbf{x}_2, \ldots, \mathbf{x}_k$ must be linearly dependent.

7. **Hint.** Use the Wronskian to test for independence.

8. **Hint.** Use a trigonometric identity to find a dependency relation.

11. **Hint.** If you have a set of vectors that includes the zero vector, such as, $\mathbf{v}_1, \mathbf{v}_2, \mathbf{v}_3, \mathbf{0}$, can you find scalars c_1, c_2, c_3, c_4, that are not all 0 such that

$$c_1\mathbf{v}_1 + c_2\mathbf{v}_2 + c_3\mathbf{v}_3 + c_4\mathbf{0} = \mathbf{0}$$

13. **Hint.** See Exercise 5.

14. See the hint in the book.

15. **Hint.** If

$$c_1\mathbf{y}_1 + c_2\mathbf{y}_2 + \cdots + c_k\mathbf{y}_k = \mathbf{0}$$

write this equation in terms of the $\mathbf{x}_i$'s and show that $c_1, c_2, \ldots, c_k$ must all be 0.

17. **Hint.** If $\mathbf{v}_1, \mathbf{v}_2, \ldots, \mathbf{v}_n$ are linearly independent there is an obvious way to choose a vector $\mathbf{v}$ that is not in $\mathrm{Span}(\mathbf{v}_2, \ldots, \mathbf{v}_n)$.

4 │ BASIS AND DIMENSION

Overview

A spanning set is minimal if its vectors are linearly independent. The elements of a minimal spanning set form the basic building blocks for the entire vector space, and consequently, we say that they form a *basis* for the vector space. The dimension of a vector space can be defined in terms of the size of the basis. Once we have chosen a basis for a vector space, we can represent all vectors as linear combinations of the basis vectors. This is crucial in many applications. Indeed, the key to solving many applied problem is picking the right basis for the vector space.

Important Concepts

1. **Basis.** The vectors $\mathbf{v}_1, \mathbf{v}_2, \ldots, \mathbf{v}_n$ form a *basis* for a vector space V if and only if
   (i) $\mathbf{v}_1, \ldots, \mathbf{v}_n$ are linearly independent.
   (ii) $\mathbf{v}_1, \ldots, \mathbf{v}_n$ span V.

2. **Dimension.** Let V be a vector space. If V has a basis consisting of n vectors, we say that V has *dimension n*. The subspace $\{\mathbf{0}\}$ of V is said to have *dimension* 0. V is said to be *finite-dimensional* if there is a finite set of vectors that spans V; otherwise, we say that V is *infinite-dimensional*.

3. **Standard basis for R^n.** The standard basis for R^n is the set $\{\mathbf{e}_1, \mathbf{e}_2, \ldots, \mathbf{e}_n\}$.

Important Theorems

▶ THEOREM 3.4.1. *If $\{\mathbf{v}_1, \mathbf{v}_2, \ldots, \mathbf{v}_n\}$ is a spanning set for a vector space V, then any collection of m vectors in V, where $m > n$, is linearly dependent.*

▶ COROLLARY 3.4.2. *If $\{\mathbf{v}_1, \ldots, \mathbf{v}_n\}$ and $\{\mathbf{u}_1, \ldots, \mathbf{u}_m\}$ are both bases for a vector space V, then $n = m$.*

▶ THEOREM 3.4.3. *If V is a vector space of dimension $n > 0$:*

(I) *Any set of n linearly independent vectors spans V.*

(II) *Any n vectors that span V are linearly independent.*

▶ THEOREM 3.4.4. *If V is a vector space of dimension $n > 0$, then:*

(i) *No set of less than n vectors can span V.*

(ii) *Any subset of less than n linearly independent vectors can be extended to form a basis for V.*

(iii) *Any spanning set containing more than n vectors can be pared down to form a basis for V.*

Exercises: Solutions and Hints

2. **Hint.** Since the dimension of R^3 is 3, a basis must consist of exactly 3 vectors and they must be linearly independent. If the set has either more or less than 3 vectors then it cannot be a basis. If the set consists of 3 vectors it will be a basis for R^3 if and only if the vectors are linearly independent.

3. (a) **Hint.** By Theorem 3.4.3, it suffices to show that the two vectors are linearly independent.

   (b) **Hint.** Apply Theorem 3.4.1.

5. (a) **Solution.** Since

$$\begin{vmatrix} 2 & 3 & 2 \\ 1 & -1 & 6 \\ 3 & 4 & 4 \end{vmatrix} = 0$$

it follows that $\mathbf{x}_1$, $\mathbf{x}_2$, $\mathbf{x}_3$ are linearly dependent.

8 (a) **Hint.** See Theorem 3.4.4.

9. (a) **Solution.** If $\mathbf{a}_1$ and $\mathbf{a}_2$ are linearly independent then they span a two dimensional subspace of R^3. A two dimensional subspace of R^3 corresponds to a plane through the origin in 3-space.

   (b) **Hint.** If $\mathbf{b} = A\mathbf{x}$ then how is $\mathbf{b}$ related to the column vectors of A?

10. **Hint.** You must find a subset of three vectors that are linearly independent.

17. **Hint.** The hint in the book should help you to prove your answer analytically. To describe the possibilities geometrically recall that a 2-dimensional subspace of R^3 corresponds to a plane through the origin in 3-space.

18. **Hint.** Show that if $\{\mathbf{u}_1, \mathbf{u}_2, \ldots, \mathbf{u}_j\}$ is a basis for U and $\{\mathbf{v}_1, \mathbf{v}_2, \ldots, \mathbf{v}_k\}$ is a basis for V then $\mathbf{u}_1, \ldots, \mathbf{u}_j, \mathbf{v}_1, \mathbf{v}_2, \ldots, \mathbf{v}_k$ are linearly independent.

5 | CHANGE OF BASIS

Overview

Many applied problems can be simplified by changing from one coordinate system to another. Changing coordinate systems in a vector space is essentially the same as switching from one basis to another. In this section we learn how to determine transition matrices which allow us to change bases by simply performing matrix multiplications.

Important Concepts

1. **Coordinate vector.** Let V be a vector space and let $E = [\mathbf{v}_1, \mathbf{v}_2, \ldots, \mathbf{v}_n]$ be an ordered basis for V. If $\mathbf{v}$ is any element of V, then $\mathbf{v}$ can be written in the form

$$\mathbf{v} = c_1\mathbf{v}_1 + c_2\mathbf{v}_2 + \cdots + c_n\mathbf{v}_n$$

where $c_1, c_2, \ldots, c_n$ are scalars. Thus we can associate with each vector $\mathbf{v}$ a unique vector $\mathbf{c} = (c_1, c_2, \ldots, c_n)^T$ in R^n. The vector $\mathbf{c}$ defined in this way is called the *coordinate vector* of $\mathbf{v}$ with respect to the ordered basis E and is denoted $[\mathbf{v}]_E$. The c_i's are called the *coordinates* of $\mathbf{v}$ relative to E.

2. **Transition matrix.** Let E and F be two ordered basis for a vector space V. For each $\mathbf{x} \in V$, let $[\mathbf{x}]_E$ denote the coordinate vector of $\mathbf{x}$ with respect to E and let $[\mathbf{x}]_F$ denote the coordinate vector of $\mathbf{x}$ with respect to F. A matrix T is said to be a *transition matrix* from E to F if

$$T[\mathbf{x}]_E = [\mathbf{x}]_F$$

for all $\mathbf{x} \in V$.

3. **Stochastic matrix.** An $n \times n$ matrix is said to be *stochastic* if its entries are all nonnegative and the entries in each column add up to 1.

Important Results

▶ If $[\mathbf{u}_1, \mathbf{u}_2, \ldots, \mathbf{u}_n]$ is an ordered basis for R^n and $U = (\mathbf{u}_1, \mathbf{u}_2, \ldots, \mathbf{u}_n)$, then U^{-1} is the transition matrix from the standard basis to $[\mathbf{u}_1, \mathbf{u}_2, \ldots, \mathbf{u}_n]$.

▶ Let $E = [\mathbf{u}_1, \mathbf{u}_2, \ldots, \mathbf{u}_n]$ and $F = [\mathbf{v}_1, \mathbf{v}_2, \ldots, \mathbf{v}_n]$ be two ordered basis for R^n and let $U = (\mathbf{u}_1, \mathbf{u}_2, \ldots, \mathbf{u}_n)$ and $V = (\mathbf{v}_1, \mathbf{v}_2, \ldots, \mathbf{v}_n)$. If $\mathbf{c}$ is the coordinate vector of $\mathbf{x}$ with respect to E and $\mathbf{d}$ is the coordinate vector of $\mathbf{x}$ with respect to F, then

$$U\mathbf{c} = \mathbf{x} = V\mathbf{d}$$

The transition matrix from E to F is $V^{-1}U$ and transition matrix from F to E is $U^{-1}V$.

Exercises: Solutions and Hints

1. (b) **Solution.** Since we are changing to the standard basis the transition matrix is just

$$U = (\mathbf{u}_1, \mathbf{u}_2) = \begin{bmatrix} 1 & 2 \\ 2 & 5 \end{bmatrix}$$

2. (b) **Solution.** Since we are changing from standard basis to the new basis $\mathbf{u}_1, \mathbf{u}_2]$ the transition matrix is

$$U^{-1} = \begin{bmatrix} 5 & -2 \\ -2 & 1 \end{bmatrix}$$

11. **Solution.** The transition matrix from E to F is $U^{-1}V$. To compute $U^{-1}V$, note that

$$U^{-1}(U \mid V) = (I \mid U^{-1}V)$$

and hence $(I \mid U^{-1}V)$ and $(U \mid V)$ are row equivalent. Thus $(I \mid U^{-1}V)$ is the reduced row echelon form of $(U \mid V)$.

6 ROW SPACE AND COLUMN SPACE

Overview

In Section 2 we learned about the nullspace of a matrix. Two other important subspaces are the row space and the column space of the matrix. In this section we examine all three of these subspaces and investigate how the dimensions of these spaces are related. The concept of the *rank* of a matrix is also introduced in this section.

Important Concepts

1. **Row space.** If A is an $m \times n$ matrix, the subspace of $R^{1 \times n}$ spanned by the row vectors of A is called the *row space* of A.
2. **Column space.** If A is an $m \times n$ matrix, the subspace of R^m spanned by the column vectors of A is called the *column space* of A.
3. **Rank of a matrix.** The *rank* of a matrix A is the dimension of the row space of A.

Important Theorems

▶ THEOREM 3.6.1. *Two row equivalent matrices have the same row space.*

▶ THEOREM 3.6.2 (**Consistency Theorem for Linear Systems**) *A linear system $A\mathbf{x} = \mathbf{b}$ is consistent if and only if $\mathbf{b}$ is in the column space of A.*

▶ THEOREM 3.6.3. *Let A be an $m \times n$ matrix. The linear system $A\mathbf{x} = \mathbf{b}$ is consistent for every $\mathbf{b} \in R^m$ if and only if the column vectors of A span R^m. The system $A\mathbf{x} = \mathbf{b}$ has at most one solution for every $\mathbf{b} \in R^m$ if and only if the column vectors of A are linearly independent.*

▶ COROLLARY 3.6.4. *An $n \times n$ matrix A is nonsingular if and only if the column vectors of A form a basis for R^n.*

▶ THEOREM 3.6.5 (**The Rank-Nullity Theorem**) *If A is an $m \times n$ matrix, then the rank of A plus the nullity of A equals n.*

▶ THEOREM 3.6.6. *If A is an $m \times n$ matrix, the dimension of the row space of A equals the dimension of the column space of A.*

Exercises: Solutions and Hints

1. (a) **Solution.** The reduced row echelon form of the matrix is

$$\begin{pmatrix} 1 & 0 & 2 \\ 0 & 1 & 0 \\ 0 & 0 & 0 \end{pmatrix}$$

Thus $(1, \ 0, \ 2)$ and $(0, \ 1, \ 0)$ form a basis for the row space. The lead 1's occur in the first and second columns, so those columns of the original matrix form a basis for the column space:

$$\mathbf{a}_1 = (1, \ 2, \ 4)^T \quad \text{and} \quad \mathbf{a}_2 = (3, \ 1, \ 7)^T$$

The reduced row echelon form involves one free variable and hence the nullspace will have dimension 1. Setting $x_3 = 1$, we get $x_1 = -2$ and $x_2 = 0$. Thus $(-2, \ 0, \ 1)^T$ is a basis for the nullspace.

3 (b) **Hint.** The column vectors of A satisfy the same dependency relations that the column vectors of U satisfy.

4. (b) **Solution.** The column space of A is the one dimensional subspace spanned by $\mathbf{a}_1$. Since $\mathbf{b}$ is not in this subspace, the system is inconsistent.

6. **Hint.** Use the consistency theorem to determine if the system is consistent. If A is an $m \times n$ matrix with linearly dependent columns, then what can you conclude about its nullity? For a consistent system what does the nullity of the coefficient matrix tell you about the number of solutions?

7. (a) **Hint.** If $N(A) = \{\mathbf{0}\}$, then what will the rank of A be?
   (b) **Hint.** Use the consistency theorem. If a system is consistent and $N(A) = \{\mathbf{0}\}$, then what can you conclude about the number of solutions? Alternatively, Theorem 3.3.2 could be helpful in answering these questions.

9. (a) **Hint.** If A and B are row equivalent, then how are their ranks related?

(b) **Hint.** Look at some examples. Make up a singular matrix A whose entries are all nonzero and then reduce A to its echelon form U. Do U and A have the same column spaces?

10. **Hint.** The column vectors of A satisfy the same dependency relations that the column vectors of U satisfy.

11. **Hint.** Show that the system must be consistent and that the echelon form of the coefficient matrix will involve free variables.

12. See the hint for Exercise 10.

14. **Hint.** $\mathbf{y} = A\mathbf{x} \neq \mathbf{0}$ is equivalent to saying that $\mathbf{x}$ is not in the nullspace of A.

16. (a) **Hint.** Show that if $\mathbf{x} \in N(A)$ then $\mathbf{x} \in N(BA)$ and show the converse, if $\mathbf{x} \in N(BA)$ then $\mathbf{x} \in N(A)$.
    (b) **Hint.** Use part (a) to show first that $(AC)^T$ and A^T have the same rank.

19. (a) **Hint.** Partition B into columns and perform the block multiplication
$$AB = A(\mathbf{b}_1, \mathbf{b}_2, \ldots, \mathbf{b}_n) = (A\mathbf{b}_1, A\mathbf{b}_2, \ldots, A\mathbf{b}_n)$$

21. (b) **Hint.** Use the Rank-Nullity theorem.

22. (a) **Hint.** Show that each column vector of C is a linear combination of the column vectors of A.
    (b) **Hint.** $C^T = B^T A^T$.

23. (a) **Hint.** In general a matrix E will have linearly independent column vectors if and only if $E\mathbf{x} = \mathbf{0}$ has only the trivial solution $\mathbf{x} = \mathbf{0}$. One way to show that C has linearly independent column vectors is to show that $C\mathbf{x} \neq \mathbf{0}$ for all $\mathbf{x} \neq \mathbf{0}$ and hence that $C\mathbf{x} = \mathbf{0}$ has only the trivial solution.
    (b) **Hint.** $C^T = B^T A^T$.

24. (a) **Hint.** If the column vectors of B are linearly dependent then $B\mathbf{x} = \mathbf{0}$ for some nonzero vector $\mathbf{x} \in R^r$.

25. (a) **Hint.** To get started let $\mathbf{b}$ be any vector in R^m and let $\mathbf{c} = C\mathbf{b}$.
    (b) **Hint.** If n vectors span R^m then how must m and n be related?

28. **Hint.** Let B be an $n \times m$ matrix. If B has a left inverse, then B^T has a right inverse. Apply the result of Exercise 25 to B^T.

30. **Solution.** Let $\mathbf{u}(1,:), \mathbf{u}(2,:), \ldots, \mathbf{u}(k,:)$ be the nonzero row vectors of U. If
$$c_1 \mathbf{u}(1,:) + c_2 \mathbf{u}(2,:) + \cdots + c_k \mathbf{u}(k,:) = \mathbf{0}^T$$
then we claim
$$c_1 = c_2 = \cdots = c_k = 0$$
This is true since the leading nonzero entry in $\mathbf{u}(i,:)$ is the only nonzero entry in its column. Let us refer to the column containing the leading nonzero entry of $\mathbf{u}(i,:)$ as $j(i)$. Thus if
$$\mathbf{y}^T = c_1 \mathbf{u}(1,:) + c_2 \mathbf{u}(2,:) + \cdots + c_k \mathbf{u}(k,:) = \mathbf{0}^T$$

then

$$0 = y_{j(i)} = c_i, \qquad i = 1, \ldots, k$$

and it follows that the nonzero row vectors of U are linearly independent.

MATLAB EXERCISES

3. (a) The column space of C is a subspace of the column space of B. Thus A and B must have the same column space and hence the same rank. Therefore we would expect the rank of A to be 4.

   (b) The first four columns of A should be linearly independent and hence should form a basis for the column space of A. The first four columns of the reduced row echelon form of A should be the same as the first four columns of the 8×8 identity matrix. Since the rank is 4, the last four rows should consist entirely of 0's.

4. (d) The vectors $C\mathbf{y}$ and $\mathbf{b} + c\mathbf{u}$ are equal since

$$C\mathbf{y} = (A + \mathbf{u}\mathbf{v}^T)\mathbf{y} = A\mathbf{y} + c\mathbf{u} = \mathbf{b} + c\mathbf{u}$$

The vectors $C\mathbf{z}$ and $(1 + d)\mathbf{u}$ are equal since

$$C\mathbf{z} = (A + \mathbf{u}\mathbf{v}^T)\mathbf{z} = A\mathbf{z} + d\mathbf{u} = \mathbf{u} + d\mathbf{u}$$

It follows that

$$C\mathbf{x} = C(\mathbf{y} - e\mathbf{z}) = \mathbf{b} + c\mathbf{u} - e(1 + d)\mathbf{u} = \mathbf{b}$$

The rank one update method will fail if $d = -1$. In this case

$$C\mathbf{z} = (1 + d)\mathbf{u} = \mathbf{0}$$

Since $\mathbf{z}$ is nonzero, the matrix C must be singular.

CHAPTER TEST A

1. **Hint.** See the discussion at the beginning of Section 2.
2. **Hint.** If $\mathbf{x}$ is a vector in R^2 then it has only 2 entries.
3. **Hint.** A two dimensional subspace of R^3 corresponds to a plane through the origin in 3-space.
4. **Hint.** See Exercise 19 of Section 2.
5. **Hint.** See Exercise 18 of Section 2.
6. **Hint.** See Theorem 3.4.3.
7. **Hint.** Is it possible for three vectors $\mathbf{x}_1, \mathbf{x}_2, \mathbf{x}_3$ to span the vector space R^2?
8. **Hint.** If

$$\text{Span}(\mathbf{x}_1, \mathbf{x}_2, \ldots, \mathbf{x}_k) = \text{Span}(\mathbf{x}_1, \mathbf{x}_2, \ldots, \mathbf{x}_{k-1})$$

   then $\mathbf{x}_k$ must be in $\text{Span}(\mathbf{x}_1, \mathbf{x}_2, \ldots, \mathbf{x}_{k-1})$.
9. **Hint.** See Theorem 3.6.5.
10. **Hint.** Use the Rank-Nullity theorem to determine the nullities of the matrices.

CHAPTER TEST B

1. **Hint.** Find a nontrivial linear combination of $\mathbf{x}_1$, $\mathbf{x}_2$, $\mathbf{x}_3$ that equals $\mathbf{0}$.

2. (b) **Hint.** S_2 consists of all vectors of the form

$$\begin{pmatrix} a \\ 0 \end{pmatrix} \quad \text{or} \quad \begin{pmatrix} 0 \\ b \end{pmatrix}$$

3. (b) **Solution.** The lead 1's in the row echelon form occur in the first and third columns, so $\mathbf{a}_1$ and $\mathbf{a}_3$ will form a basis for the column space of A and the rank of A will be 2..

5. **Hint.** One dimensional subspaces correspond to lines through the origin and two dimensional correspond subspaces correspond to planes through the origin.

6. **Hint.** The vectors in S are matrices of the form

$$\begin{pmatrix} a & b \\ b & c \end{pmatrix}$$

7. (c) **Hint.** The consistency theorem guarantees that there will be at least one solution. Whether or not there are more depends upon whether there are free variables in the echelon form of the coefficient matrix.

9. **Hint.** You should be able to show linear independence by just working from the definition.

10. (a) **Hint. Hint.** First determine the rank of A.
    (b) **Hint.** If A and U are row equivalent matrices then their column vectors satisfy the same dependency relations.

11. (b) **Hint.** The equation $c_1 \mathbf{v}_1 + c_2 \mathbf{v}_2 = d_1 \mathbf{u}_1 + d_2 \mathbf{u}_2$ can be rewritten as a matrix equation $V\mathbf{c} = U\mathbf{d}$.

Chapter 4

Linear

ormations

map one vector space into another and
that

$$L(\mathbf{v}_1) + c_2 L(\mathbf{v}_2) + \cdots + c_n L(\mathbf{v}_n)$$

ations. We look at a number of examples
omain and range that are associated with

Important Concepts

1. **Linear transformation.** A mapping L from a vector space V into a vector space W is said to be a *linear transformation* if
$$L(\alpha\mathbf{v}_1 + \beta\mathbf{v}_2) = \alpha L(\mathbf{v}_1) + \beta L(\mathbf{v}_2)$$
for all $\mathbf{v}_1, \mathbf{v}_2 \in V$ and for all scalars α and β.

2. **Linear Operator.** If L is a linear transformation mapping a vector space V into itself then we say that L is a linear operator on V.

3. **Kernel of a linear transformation.** Let $L\colon V \to W$ be a linear transformation. The *kernel* of L, denoted $\ker(L)$, is defined by
$$\ker(L) = \{\mathbf{v} \in V | L(\mathbf{v}) = \mathbf{0}_W\}$$

4. **Image of a linear transformation.** Let $L\colon V \to W$ be a linear transformation and let S be a subspace of V. The *image* of S, denoted $L(S)$, is defined by
$$L(S) = \{\mathbf{w} \in W | \mathbf{w} = L(\mathbf{v}) \quad \text{for some} \quad \mathbf{v} \in S\}$$
The image of the entire vector space, $L(V)$, is called the *range* of L.

Important Theorems

▶ **THEOREM 4.1.1.** *If $L\colon V \to W$ is a linear transformation and S is a subspace of V, then*

(i) $\ker(L)$ *is a subspace of V.*
(ii) $L(S)$ *is a subspace of W.*

Exercises: Solutions and Hints

2. **Solution.** The polar coordinate representation is
$$x_1 = r\cos\theta \quad \text{and} \quad x_2 = r\sin\theta$$
where $r = (x_1^2 + x_2^2)^{1/2}$ and θ is the angle between $\mathbf{x}$ and $\mathbf{e}_1$.
$$\begin{aligned}L(\mathbf{x}) &= (r\cos\theta\cos\alpha - r\sin\theta\sin\alpha, r\cos\theta\sin\alpha + r\sin\theta\cos\alpha)^T \\ &= (r\cos(\theta+\alpha), r\sin(\theta+\alpha))^T\end{aligned}$$
The linear transformation L has the effect of rotating a vector by an α in the counterclockwise direction.

8. (b) **Solution.**
$$L(\alpha A + \beta B) = C^2(\alpha A + \beta B) = \alpha C^2 A + \beta C^2 B = \alpha L(A) + \beta L(B)$$
Therefore L is a linear operator.

10. **Solution.** If $f, g \in C[0,1]$ then
$$L(\alpha f + \beta g) = \int_0^x (\alpha f(t) + \beta g(t))dt$$

$$= \alpha \int_0^x f(t)dt + \beta \int_0^x g(t)dt$$
$$= \alpha L(f) + \beta L(g)$$

Thus L is a linear transformation from $C[0,1]$ to $C[0,1]$.

13. Hint. If $\mathbf{v}$ is any element of V then

$$\mathbf{v} = \alpha_1 \mathbf{v}_1 + \alpha_2 \mathbf{v}_2 + \cdots + \alpha_n \mathbf{v}_n$$

15. Solution. The proof is by induction on n. In the case $n = 1$, L^1 is a linear operator since $L^1 = L$. We will show that if L^m is a linear operator on V then L^{m+1} is also a linear operator on V. This follows since

$$L^{m+1}(\alpha \mathbf{v}) = L(L^m(\alpha \mathbf{v})) = L(\alpha L^m(\mathbf{v})) = \alpha L(L^m(\mathbf{v})) = \alpha L^{m+1}(\mathbf{v})$$

and

$$\begin{aligned} L^{m+1}(\mathbf{v}_1 + \mathbf{v}_2) &= L(L^m(\mathbf{v}_1 + \mathbf{v}_2)) \\ &= L(L^m(\mathbf{v}_1) + L^m(\mathbf{v}_2)) \\ &= L(L^m(\mathbf{v}_1)) + L(L^m(\mathbf{v}_2)) \\ &= L^{m+1}(\mathbf{v}_1) + L^{m+1}(\mathbf{v}_2) \end{aligned}$$

17. (b) Solution. Since

$$L(\mathbf{x}) = \begin{pmatrix} x_1 \\ x_2 \\ 0 \end{pmatrix} = x_1 \mathbf{e}_1 + x_2 \mathbf{e}_2$$

it follows that the range of L is $\mathrm{Span}(\mathbf{e}_1, \mathbf{e}_2)$. A vector $\mathbf{x}$ is in the kernel of L if and only if $x_1 = x_2 = 0$. Therefore

$$\ker(L) = \mathrm{Span}(\mathbf{e}_3)$$

18. (b) Solution. L is the identity operator on S, so

$$L(S) = S \qquad \text{and} \qquad \ker(L) = \{\mathbf{0}\}$$

20. Hint. Show that if $\mathbf{v} \in L^{-1}(T)$, then $L(c\mathbf{v}) \in T$ for any scalar c and show that if $\mathbf{v}$ and $\mathbf{w}$ are in $L^{-1}(T)$, then $L(\mathbf{v} + \mathbf{w}) \in T$.

21. Hint. Show that if L is one-to-one and $\mathbf{v} \in \ker(L)$, then $\mathbf{v}$ must be the zero vector. Show that if $\ker(L) = \{\mathbf{0}\}$ and $L(\mathbf{v}_1) = L(\mathbf{v}_2)$ then $\mathbf{v}_1$ must equal $\mathbf{v}_2$.

25. Solution.
(a) If $p = ax^2 + bx + c \in P_3$, then

$$D(p) = 2ax + b$$

Thus

$$D(P_3) = \mathrm{Span}(1, x) = P_2$$

The operator is not one-to-one, for if $p_1(x) = ax^2 + bx + c_1$ and $p_2(x) = ax^2 + bx + c_2$ where $c_2 \neq c_1$, then $D(p_1) = D(p_2)$.

(b) The subspace S consists of all polynomials of the form $ax^2 + bx$. If $p_1 = a_1x^2 + b_1x$, $p_2 = a_2x^2 + b_2x$ and $D(p_1) = D(p_2)$, then

$$2a_1x + b_1 = 2a_2x + b_2$$

and it follows that $a_1 = a_2$, $b_1 = b_2$. Thus $p_1 = p_2$ and hence D is one-to-one. D does not map S onto P_3 since $D(S) = P_2$.

2 | MATRIX REPRESENTATIONS OF LINEAR TRANSFORMATIONS

Overview

Every linear transformation from R^n to R^m can be viewed as a matrix vector multiplication. Indeed, if $L : R^n \to R^m$, then there exists an $m \times n$ matrix A such that $L(\mathbf{x}) = A\mathbf{x}$ for each $\mathbf{x} \in R^n$. More generally, if V is an n-dimensional vector space and W is an m-dimensional vector space and $L : V \to W$, then if we specify ordered bases for V and W it is possible to represent the action of L with respect to those bases by an $m \times n$ matrix.

Important Concepts

1. **Standard matrix representation.** If L is a linear transformation mapping R^n into R^m, then there exists an $m \times n$ matrix A such that $L(\mathbf{x}) = A\mathbf{x}$ for each $\mathbf{x} \in R^n$. The matrix A is said to be the *standard matrix representation* of L.

2. **Coordinate vector.** If $E = [\mathbf{v}_1, \mathbf{v}_2, , \ldots, \mathbf{v}_n]$ is an ordered basis for a vector space V and $\mathbf{v} \in V$, then $\mathbf{v}$ can be represented as a linear combination

$$\mathbf{v} = c_1\mathbf{v}_1 + c_2\mathbf{v}_2 + \cdots + c_n\mathbf{v}_n$$

The vector of scalars $\mathbf{c} = (c_1, c_2, \ldots, c_n)^T$ is said to be the *coordinate vector* of $\mathbf{v}$ with respect to E. It is denoted by $[\mathbf{v}]_E$.

3. **General matrix representation.** Let $E = [\mathbf{v}_1, \mathbf{v}_2, , \ldots, \mathbf{v}_n]$ be an ordered basis for a vector space V and let $F = [\mathbf{u}_1, \mathbf{u}_2, , \ldots, \mathbf{u}_m]$ be an ordered basis for a vector space W. If L is a linear transformation mapping V into W, then there exists an $m \times n$ matrix A such that $[L(\mathbf{v})]_F = A[\mathbf{v}]_E$ for each $\mathbf{x} \in R^n$. The matrix A is said to be the *matrix representation* of L with respect to the ordered bases E and F. Thus if $\mathbf{c} = [\mathbf{v}]_E$ and $\mathbf{d} = [L(\mathbf{v})]_F$ and A is the matrix representation of L, then

$$\mathbf{v} = c_1\mathbf{v}_1 + c_2\mathbf{v}_2 + \cdots + c_n\mathbf{v}_n$$

and

$$L(\mathbf{v}) = d_1\mathbf{u}_1 + d_2\mathbf{u}_2 + \cdots + d_m\mathbf{u}_m$$

where $\mathbf{d}$ is determined by the matrix multiplication

$$\mathbf{d} = A\mathbf{c}$$

4. Homogeneous Coordinate System. The *homogeneous coordinate system* is formed by equating each vector in R^2 with a vector in R^3 having the same first two coordinates and having 1 as it third coordinate.

$$\begin{pmatrix} x_1 \\ x_2 \end{pmatrix} \leftrightarrow \begin{pmatrix} x_1 \\ x_2 \\ 1 \end{pmatrix}$$

Using the homogeneous coordinate system we are able to treat translations in the plane as linear transformations and represent them by 3×3 matrices.

Important Theorems

▶ THEOREM 4.2.1. *If L is a linear operator mapping R^n into R^m, there is an $m \times n$ matrix A such that*

$$L(\mathbf{x}) = A\mathbf{x}$$

for each $\mathbf{x} \in R^n$. In fact, the jth column vector of A is given by

$$\mathbf{a}_j = L(\mathbf{e}_j) \qquad j = 1, 2, \ldots, n$$

▶ THEOREM 4.2.2 **(Matrix Representation Theorem)** *If $E = [\mathbf{v}_1, \mathbf{v}_2, \ldots, \mathbf{v}_n]$ and $F = [\mathbf{w}_1, \mathbf{w}_2, \ldots, \mathbf{w}_m]$ are ordered bases for vector spaces V and W, respectively, then corresponding to each linear transformation $L : V \to W$ there is an $m \times n$ matrix A such that*

$$[L(\mathbf{v})]_F = A[\mathbf{v}]_E \qquad \text{for each } \mathbf{v} \in V$$

A is the matrix representing L relative to the ordered bases E and F. In fact,

$$\mathbf{a}_j = [L(\mathbf{v}_j)]_F \qquad j = 1, 2, \ldots, n$$

▶ THEOREM 4.2.3. *Let $E = [\mathbf{u}_1, \ldots, \mathbf{u}_n]$ and $F = [\mathbf{b}_1, \ldots, \mathbf{b}_m]$ be ordered bases for R^n and R^m, respectively. If $L : R^n \to R^m$ is a linear transformation and A is the matrix representing L with respect to E and F, then*

$$\mathbf{a}_j = B^{-1}L(\mathbf{u}_j) \qquad \text{for } j = 1, \ldots, n$$

where $B = (\mathbf{b}_1, \ldots, \mathbf{b}_m)$.

▶ COROLLARY 4.2.4. *If A is the matrix representing the linear operator $L : R^n \to R^m$ with respect to the bases*

$$E = [\mathbf{u}_1, \ldots, \mathbf{u}_n] \quad and \quad F = [\mathbf{b}_1, \ldots, \mathbf{b}_m]$$

then the reduced row echelon form of $(\mathbf{b}_1, \ldots, \mathbf{b}_m \mid L(\mathbf{u}_1), \ldots, L(\mathbf{u}_n))$ is $(I \mid A)$.

Exercises: Solutions and Hints

3. (b) **Solution.** To determine the columns of A we apply L to each of the standard basis vectors for R^3.

$$\mathbf{a}_1 = L(\mathbf{e}_1) = \begin{pmatrix} 1 \\ 1 \\ 1 \end{pmatrix}, \qquad \mathbf{a}_2 = L(\mathbf{e}_2) = \begin{pmatrix} 0 \\ 1 \\ 1 \end{pmatrix}, \qquad \mathbf{a}_3 = L(\mathbf{e}_3) = \begin{pmatrix} 0 \\ 0 \\ 1 \end{pmatrix}$$

Therefore

$$A = (\mathbf{a}_1, \mathbf{a}_2, \mathbf{a}_3) = \begin{pmatrix} 1 & 0 & 0 \\ 1 & 1 & 0 \\ 1 & 1 & 1 \end{pmatrix}$$

5. (c) **Solution.** To determine the standard matrix representation of L we look at its effect on the standard basis vector $\mathbf{e}_1$ and $\mathbf{e}_2$.

$$\mathbf{a}_1 = L(\mathbf{e}_1) = \begin{pmatrix} 2\cos\frac{\pi}{6} \\ 2\sin\frac{\pi}{6} \end{pmatrix} = \begin{pmatrix} \sqrt{3} \\ 1 \end{pmatrix}$$

$$\mathbf{a}_2 = L(\mathbf{e}_2) = \begin{pmatrix} 2\cos\frac{2\pi}{3} \\ 2\sin\frac{2\pi}{3} \end{pmatrix} = \begin{pmatrix} -1 \\ \sqrt{3} \end{pmatrix}$$

The matrix representation is

$$A = (\mathbf{a}_1, \mathbf{a}_2) = \begin{pmatrix} \sqrt{3} & -1 \\ 1 & \sqrt{3} \end{pmatrix}$$

7. Hint. In part (a) we must a vector $\mathbf{c} = (c_1, c_2, c_3)^T$ such that

$$\mathcal{I}(\mathbf{e}_1) = c_1\mathbf{y}_1 + c_2\mathbf{y}_2 + c_3\mathbf{y}_3$$

Since $\mathcal{I}$ is the identity operator, $\mathcal{I}(\mathbf{e}_1) = \mathbf{e}_1$, and hence we can rewrite the equation in the form

$$Y\mathbf{c} = \mathbf{e}_1$$

It follows that

$$\mathbf{c} = Y^{-1}\mathbf{e}_1$$

Note that $\mathbf{c}$ is the first column of Y^{-1} which is the transition matrix from the standard basis $[\mathbf{e}_1, \mathbf{e}_2, \mathbf{e}_3]$ to $[\mathbf{y}_1, \mathbf{y}_2, \mathbf{y}_3]$.

11. (c) **Solution.** The matrices for the pitch and roll are

$$P = \begin{pmatrix} \cos\frac{\pi}{4} & 0 & -\sin\frac{\pi}{4} \\ 0 & 1 & 0 \\ \sin\frac{\pi}{4} & 0 & \cos\frac{\pi}{4} \end{pmatrix} \quad \text{and} \quad R = \begin{pmatrix} 1 & 0 & 0 \\ 0 & \cos\frac{-\pi}{2} & -\sin\frac{-\pi}{2} \\ 0 & \sin\frac{-\pi}{2} & \cos\frac{-\pi}{2} \end{pmatrix}$$

The matrix for the combined transformation is

$$Q = PR = \begin{pmatrix} \frac{1}{\sqrt{2}} & 0 & -\frac{1}{\sqrt{2}} \\ 0 & 1 & 0 \\ \frac{1}{\sqrt{2}} & 0 & \frac{1}{\sqrt{2}} \end{pmatrix} \begin{pmatrix} 1 & 0 & 0 \\ 0 & 0 & 1 \\ 0 & -1 & 0 \end{pmatrix} = \begin{pmatrix} \frac{1}{\sqrt{2}} & \frac{1}{\sqrt{2}} & 0 \\ 0 & 0 & 1 \\ \frac{1}{\sqrt{2}} & -\frac{1}{\sqrt{2}} & 0 \end{pmatrix}$$

14. Partial Solution. To determine the matrix representation we must represent $L(x^2)$, $L(x)$ and $L(1)$ in terms of the ordered basis $[2, 1-x]$.

$$L(x^2) = 2x = 1 \cdot 2 - 2(1-x)$$
$$L(x) = 1 = \frac{1}{2} \cdot 2 + 0(1-x)$$
$$L(1) = 1 = \frac{1}{2} \cdot 2 + 0(1-x)$$

So the matrix representation is

$$A = \begin{pmatrix} 1 & \frac{1}{2} & \frac{1}{2} \\ -2 & 0 & 0 \end{pmatrix}$$

16. Hint. Make use of Theorem 1.4.2.

18. (b) **Hint.**

$$L(\mathbf{u}_1) = \begin{pmatrix} 0 \\ 2 \end{pmatrix}, \quad L(\mathbf{u}_2) = \begin{pmatrix} 2 \\ 0 \end{pmatrix}, \quad L(\mathbf{u}_1) = \begin{pmatrix} 0 \\ -2 \end{pmatrix}$$

Use Corollary 4.2.4 to find the matrix representation of L.

20. (a) **Solution.** If A is the matrix representing L with respect to E and F, it follows that $L(\mathbf{v}) = \mathbf{0}_W$ if and only if $A[\mathbf{v}]_E = \mathbf{0}$. Thus $\mathbf{v} \in \ker(L)$ if and only if $[\mathbf{v}]_E \in N(A)$.

3 | SIMILARITY

Overview

If L is a linear transformation mapping an n-dimensional vector space V into itself, the matrix representation of L will depend on the ordered basis chosen for V. By using different bases, it is possible to represent L by different $n \times n$ matrices. In this section we consider different matrix representations of linear operators and characterize the relationship between matrices representing the same linear operator.

Important Concepts

1. Similar matrices. Let A and B be $n \times n$ matrices. B is said to be *similar* to A if there exists a nonsingular matrix S such that $B = S^{-1}AS$.

Important Theorems

▶ THEOREM 4.3.1. *Let* $E = [\mathbf{v}_1, \ldots, \mathbf{v}_n]$ *and* $F = [\mathbf{w}_1, \ldots, \mathbf{w}_n]$ *be two ordered bases for a vector space* V, *and let* L *be a linear operator on* V. *Let* S *be the transition matrix representing the change from* F *to* E. *If* A *is the matrix representing* L *with respect to* E *and* B *is the matrix representing* L *with respect to* F, *then* $B = S^{-1}AS$.

Exercises: Solutions and Hints

4. Solution. The transition matrix from $[\mathbf{v}_1, \mathbf{v}_2, \mathbf{v}_3]$ to the standard basis $[\mathbf{e}_1, \mathbf{e}_2, \mathbf{e}_3]$ is $V = (\mathbf{v}_1, \mathbf{v}_2, \mathbf{v}_3)$ and the transition matrix from $[\mathbf{e}_1, \mathbf{e}_2, \mathbf{e}_3]$ to $[\mathbf{v}_1, \mathbf{v}_2, \mathbf{v}_3]$ is V^{-1}. Since A is the standard matrix representation of L, the matrix representation with respect to $[\mathbf{v}_1, \mathbf{v}_2, \mathbf{v}_3]$ is

$$
B = V^{-1}AV = \begin{pmatrix} -2 & 1 & 2 \\ 3 & -1 & -2 \\ 2 & -1 & -1 \end{pmatrix} \begin{pmatrix} 3 & -1 & -2 \\ 2 & 0 & -2 \\ 2 & -1 & -1 \end{pmatrix} \begin{pmatrix} 1 & 1 & 0 \\ 1 & 2 & -2 \\ 1 & 0 & 1 \end{pmatrix}
$$

$$
= \begin{pmatrix} -2 & 1 & 2 \\ 3 & -1 & -2 \\ 2 & -1 & -1 \end{pmatrix} \begin{pmatrix} 0 & 1 & 0 \\ 0 & 2 & -2 \\ 0 & 0 & 1 \end{pmatrix}
$$

$$
= \begin{pmatrix} 0 & 0 & 0 \\ 0 & 1 & 0 \\ 0 & 0 & 1 \end{pmatrix}
$$

7. Hint. If A is similar to B then there exists a nonsingular matrix S_1 such that $A = S_1^{-1}BS_1$ and if B is similar to C then there exists a nonsingular matrix S_2 such that $B = S_2^{-1}CS_2$.

8. (a) **Solution.** If $A = S\Lambda S^{-1}$, then $AS = \Lambda S$. If $\mathbf{s}_i$ is the ith column of S then $A\mathbf{s}_i$ is the ith column of AS. The ith column of ΛS is $\lambda_i \mathbf{s}_i$. Since

$AS = \Lambda S$ it follows that

$$As_i = \lambda_i s_i, \qquad i = 1, \ldots, n$$

9. **Hint.** Can you use S and S^{-1} to transform A to B?

10. **Hint.** If A is similar to B then there is a nonsingular matrix S such that

$$A = SBS^{-1}$$

If A is also equal to ST, then how must T be chosen?

11. **Hint.** If B is similar to A, then $B = S^{-1}AS$. Take determinants of both sides of this equation and make use of results from Section 2 of Chapter 2.

14. (a) **Hint.** If A and B are similar, then there exists a nonsingular matrix S such that $B = SAS^{-1}$. Look at $S(A - \lambda I)S^{-1}$.

15. (a) **Solution.** Let $C = AB$ and $E = BA$. The diagonal entries of C and E are given by

$$c_{ii} = \sum_{k=1}^{n} a_{ik}b_{ki}, \qquad e_{kk} = \sum_{i=1}^{n} b_{ki}a_{ik}$$

Hence it follows that

$$\operatorname{tr}(AB) = \sum_{i=1}^{n} c_{ii} = \sum_{i=1}^{n}\sum_{k=1}^{n} a_{ik}b_{ki} = \sum_{k=1}^{n}\sum_{i=1}^{n} b_{ki}a_{ik} = \sum_{k=1}^{n} e_{kk} = \operatorname{tr}(BA)$$

(b) **Hint.** Try to make use of the result from part (a).

MATLAB EXERCISES

2. (a) To determine the matrix representation of L with respect to E set

$$B = U^{-1}AU$$

(b) To determine the matrix representation of L with respect to F set

$$C = V^{-1}AV$$

(c) If B and C are both similar to A, then they must be similar to each other. Indeed the transition matrix S from F to E is given by $S = U^{-1}V$ and

$$C = S^{-1}BS$$

CHAPTER TEST A

1. **Hint.** If L is represented by a matrix A, then $L(\mathbf{x}) = A\mathbf{x}$ for each $\mathbf{x} \in R^2$. Does $A\mathbf{x}_1 = A\mathbf{x}_2$ necessarily imply that $\mathbf{x}_1 = \mathbf{x}_2$?

2. **Hint.** Show that

$$(L_1 + L_2)(c_1\mathbf{v}_1 + c_2\mathbf{v}_2) = c_1(L_1 + L_2)(\mathbf{v}_1) + c_2(L_1 + L_2)(\mathbf{v}_2)$$

4. **Hint.** Make up some vectors. For each vector $\mathbf{x}$ you make up, test to see if $L_1(\mathbf{x}) = L_2(\mathbf{x})$.

5. **Hint.** Check to see if the homogeneous coordinate system is closed under the operations of scalar multiplication and vector addition.

6. **Hint.** Check to see if

$$L^2(c_1\mathbf{x}_1 + c_2\mathbf{x}_2) = c_1 L^2(\mathbf{x}_1) + c_2 L^2(\mathbf{x}_2)$$

7. **Hint.** Let L_1 and L_2 be linear transformations that are both represented by the same matrix A with respect to the ordered basis E. Will $L_1(\mathbf{x}) = L_2(\mathbf{x})$ for every $\mathbf{x} \in R^n$?

8. **Hint.** See Theorem 4.3.1.

9. **Hint.** If

$$A = X^{-1}BX \qquad \text{and} \qquad B = Y^{-1}CY$$

can you come up with a matrix Z such that $A = Z^{-1}CZ$?

10. **Hint.** Can you make up an example of a singular matrix and a nonsingular matrix that have the same trace? Can a singular matrix be similar to a nonsingular matrix?

CHAPTER TEST B

1. (b) **Solution.** If $\mathbf{x} = (1,1)^T$ then $L(\mathbf{x}) = (1,1)^T$, however, $2\mathbf{x} = (2,2)^T$ and $L(2\mathbf{x}) = (4,2)^T \neq 2L(\mathbf{x})$, so L is not a linear operator.

2. **Hint.** First write $\mathbf{v}_3$ as a linear combination of $\mathbf{v}_1$ and $\mathbf{v}_2$.

3. (b) **Solution.** If $\mathbf{x}$ is in S then it must be of the form $\mathbf{x} = (c, 0, c)^T$ for some scalar c. It follows that

$$L(\mathbf{x}) = \begin{pmatrix} -c \\ c \\ 0 \end{pmatrix}$$

Therefore $L(S) = \text{Span}((-1, 1, 0)^T)$.

4. **Hint.** The range of L is a 2-dimensional subspace of R^3. Find the two basis vectors.

5. **Hint.** To find the matrix representation A determine the effect of L on $\mathbf{e}_1$ and $\mathbf{e}_2$.

6. See the hint for question 5.

7. **Solution.**

$$A = \begin{pmatrix} 1 & 0 & 2 \\ 0 & 1 & 5 \\ 0 & 0 & 1 \end{pmatrix}$$

8. **Hint.** The standard matrix representation for L is

$$A = \begin{pmatrix} \frac{1}{\sqrt{2}} & -\frac{1}{\sqrt{2}} \\ \frac{1}{\sqrt{2}} & \frac{1}{\sqrt{2}} \end{pmatrix}$$

Use $U = (\mathbf{u}_1, \mathbf{u}_2)$ and A to determine the matrix representation of L with respect to $[\mathbf{u}_1, \mathbf{u}_2]$.

10. (b) **Hint.** $I = S^{-1}IS$.

Chapter 5

Orthogonality

1 | THE SCALAR PRODUCT IN R^N

Overview

One of the main concepts in this course is that of a scalar product. In this section we see how the scalar product in R^n can be used to define lengths of vectors, angles between vectors, distance between vectors, and important statistical concepts such as correlations and covariances. Of particular importance in applications is the case where the scalar product of two vectors is 0. In this case we say that the vectors are *orthogonal*. The concept of orthogonality is the key to doing projections and to solving least squares problems.

Important Concepts

1. **Scalar Product.** If $\mathbf{x}$ and $\mathbf{y}$ are vectors in R^n, then the product

$$\mathbf{x}^T\mathbf{y} = x_1y_1 + x_2y_2 + \cdots + x_ny_n$$

is called the *scalar product* of $\mathbf{x}$ and $\mathbf{y}$.

67

2. **Length of a vector.** If $\mathbf{x}$ is a vector in R^n, then the *length* of $\mathbf{x}$, denoted $\|\mathbf{x}\|$, is defined by

$$\|\mathbf{x}\| = \sqrt{\mathbf{x}^T\mathbf{x}} = (x_1^2 + x_2^2 + \cdots + x_n^2)^{\frac{1}{2}}$$

3. **Distance between vectors.** Let $\mathbf{x}$ and $\mathbf{y}$ be vectors in R^n. The *distance between* $\mathbf{x}$ *and* $\mathbf{y}$ is defined to be the number $\|\mathbf{x} - \mathbf{y}\|$.

4. **Angle between vectors.** The *angle θ between two vectors* $\mathbf{x}$ and $\mathbf{y}$ in R^n is defined by

$$\cos\theta = \frac{\mathbf{x}^T\mathbf{y}}{\|\mathbf{x}\|\,\|\mathbf{y}\|}$$

where $0 \leq \theta \leq \pi$.

5. **Orthogonal vectors.** The vectors $\mathbf{x}$ and $\mathbf{y}$ in R^n are said to be *orthogonal* if $\mathbf{x}^T\mathbf{y} = 0$.

6. **Vector and scalar projections.** If

$$\alpha = \frac{\mathbf{x}^T\mathbf{y}}{\|\mathbf{y}\|}$$

and

$$\mathbf{p} = \alpha\frac{1}{\|\mathbf{y}\|}\mathbf{y} = \frac{\mathbf{x}^T\mathbf{y}}{\mathbf{y}^T\mathbf{y}}\mathbf{y}$$

then α is said to be the *scalar projection* of $\mathbf{x}$ onto $\mathbf{y}$, and $\mathbf{p}$ is said to be the *vector projection* of $\mathbf{x}$ onto $\mathbf{y}$.

7. **Covariance and covariance matrices.** Given a collection of n data points representing values of some variable x, we compute the mean $\overline{x}$ of the data points and form a vector $\mathbf{x}$ of the deviations from the mean. The *variance*, s^2, is defined by

$$s^2 = \frac{1}{n-1}\sum_{1}^{n} x_i^2 = \frac{\mathbf{x}^T\mathbf{x}}{n-1}$$

and the standard deviation s is the square root of the variance. If we have two data sets X_1 and X_2 each containing n values of a variable, we can form vectors $\mathbf{x}_1$ and $\mathbf{x}_2$ of deviations from the mean for both sets. The *covariance* is defined by

$$\mathrm{cov}(X_1, X_2) = \frac{\mathbf{x}_1^T\mathbf{x}_2}{n-1}$$

If we have more than two data sets, we can form a matrix X whose columns represent the deviations from the mean for each data set and then form a *covariance matrix S* by setting

$$S = \frac{1}{n-1}X^T X$$

8. **Correlation and correlation matrices.** Given two vectors X_1 and X_2 of data, to determine the *correlation* between the data sets we compute the mean value for each of the data sets and then form vectors $\mathbf{x}_1$ and $\mathbf{x}_2$ of

deviations from the mean. If we normalize $\mathbf{x}_1$ and $\mathbf{x}_2$ to make them unit vectors by setting

$$\mathbf{u}_1 = \frac{1}{\|\mathbf{x}_1\|} \quad \text{and} \quad \mathbf{u}_2 = \frac{1}{\|\mathbf{x}_2\|}$$

then the *correlation* between the two data sets is defined to be the scalar product $\mathbf{u}_1^T \mathbf{u}_2$.

Given a collection of n data vectors $X_1, X_2, \ldots, X_n$, one can define a *correlation matrix* C whose (i, j) entry is the correlation between X_i and X_j. Since the correlation is a scalar product of the form $\mathbf{u}_i^T \mathbf{u}_j$ where $\mathbf{u}_i$ and $\mathbf{u}_j$ normalized vectors of deviations from the mean, the correlation matrix C is of the form $C = U^T U$.

Important Theorems and Results

▶ THEOREM 5.1.1. *If $\mathbf{x}$ and $\mathbf{y}$ are two nonzero vectors in either R^2 or R^3 and θ is the angle between them, then*

$$\mathbf{x}^T \mathbf{y} = \|\mathbf{x}\| \, \|\mathbf{y}\| \cos \theta$$

▶ THEOREM 5.1.2 **(Cauchy–Schwarz Inequality)** *If $\mathbf{x}$ and $\mathbf{y}$ are vectors in R^n, then*

$$|\mathbf{x}^T \mathbf{y}| \leq \|\mathbf{x}\| \, \|\mathbf{y}\|$$

with equality holding if and only if one of the vectors is $\mathbf{0}$ or one vector is a multiple of the other.

▶ Pythagorean Law   If $\mathbf{x}$ and $\mathbf{y}$ are orthogonal vectors in R^n, then

$$\|\mathbf{x} + \mathbf{y}\|^2 = \|\mathbf{x}\|^2 + \|\mathbf{y}\|^2$$

Exercises: Solutions and Hints

1. (c) **Solution.** $\cos \theta = \dfrac{14}{\sqrt{221}}, \quad \theta \approx 10.65°$

3. (b) **Solution.**

$$\mathbf{p} = \frac{\mathbf{x}^T \mathbf{y}}{\mathbf{y}^T \mathbf{y}} \mathbf{y} = \frac{8}{2} \mathbf{y} = \begin{pmatrix} 4 \\ 4 \end{pmatrix}$$

It follows that

$$\mathbf{x} - \mathbf{p} = (-1, 1)^T \quad \text{and} \quad \mathbf{p}^T (\mathbf{x} - \mathbf{p}) = 4 \cdot (-1) + 4 \cdot 1 = 0$$

8. (b) **Solution.** $-3(x - 4) + 6(y - 2) + 2(z + 5) = 0$

12. **Hint.** Expand the right hand side of the equation

$$\|\mathbf{u} + \mathbf{v}\|^2 = (\mathbf{u} + \mathbf{v})^T (\mathbf{u} + \mathbf{v})$$

14. **Solution.**

(a) By the Pythagorean Theorem

$$\alpha^2 + h^2 = \|\mathbf{a}_1\|^2$$

where α is the scalar projection of $\mathbf{a}_1$ onto $\mathbf{a}_2$. It follows that

$$\alpha^2 = \frac{(\mathbf{a}_1^T \mathbf{a}_2)^2}{\|\mathbf{a}_2\|^2}$$

and

$$h^2 = \|\mathbf{a}_1\|^2 - \frac{(\mathbf{a}_1^T \mathbf{a}_2)^2}{\|\mathbf{a}_2\|^2}$$

Hence

$$h^2\|\mathbf{a}_2\|^2 = \|\mathbf{a}_1\|^2 \|\mathbf{a}_2\|^2 - (\mathbf{a}_1^T \mathbf{a}_2)^2$$

(b) If $\mathbf{a}_1 = (a_{11}, a_{21})^T$ and $\mathbf{a}_2 = (a_{12}, a_{22})^T$, then by part (a)

$$\begin{aligned} h^2\|\mathbf{a}_2\|^2 &= (a_{11}^2 + a_{21}^2)(a_{12}^2 + a_{22}^2) - (a_{11}a_{12} + a_{21}a_{22})^2 \\ &= (a_{11}^2 a_{22}^2 - 2a_{11}a_{22}a_{12}a_{21} + a_{21}^2 a_{12}^2) \\ &= (a_{11}a_{22} - a_{21}a_{12})^2 \end{aligned}$$

Therefore

$$\text{Area of } P = h\|\mathbf{a}_2\| = |a_{11}a_{22} - a_{21}a_{12}| = |\det(A)|$$

15. (a) **Solution.** If *theta* is the angle between the vectors then

$$\cos\theta = \frac{\mathbf{x}^T\mathbf{y}}{\|\mathbf{x}\|\,\|\mathbf{y}\|} = \frac{20}{8 \cdot 5} = \frac{1}{2}$$

Thus the angle between the vectors is $\theta = \frac{\pi}{3}$,

16. **Solution.**

(a) Let

$$\alpha = \frac{\mathbf{x}^T\mathbf{y}}{\mathbf{y}^T\mathbf{y}} \qquad \text{and} \qquad \beta = \frac{(\mathbf{x}^T\mathbf{y})^2}{\mathbf{y}^T\mathbf{y}}$$

In terms of these scalars we have $\mathbf{p} = \alpha\mathbf{y}$ and $\mathbf{p}^T\mathbf{x} = \beta$. Furthermore

$$\mathbf{p}^T\mathbf{p} = \alpha^2\mathbf{y}^T\mathbf{y} = \beta$$

and hence

$$\mathbf{p}^T\mathbf{z} = \mathbf{p}^T\mathbf{x} - \mathbf{p}^T\mathbf{p} = \beta - \beta = 0$$

(b) If $\|\mathbf{p}\| = 6$ and $\|\mathbf{z}\| = 8$, then we can apply the Pythagorean law to determine the length of $\mathbf{x} = \mathbf{p} + \mathbf{z}$. It follows that

$$\|\mathbf{x}\|^2 = \|\mathbf{p}\|^2 + \|\mathbf{z}\|^2 = 36 + 64 = 100$$

and hence $\|\mathbf{x}\| = 10$.

2 | ORTHOGONAL SUBSPACES

Overview

In this section we introduce the concepts of orthogonal subspaces and orthogonal complements. Each subspace S of R^m has a unique orthogonal complement, denoted $S^\perp$, which is also a subspace of R^m. Furthermore any vector $\mathbf{b}$ in R^m can be uniquely expressed as a sum $\mathbf{p} + \mathbf{z}$ where $\mathbf{p} \in S$ and $\mathbf{z} \in S^\perp$. The component $\mathbf{p}$ is the closest vector in S to $\mathbf{b}$ and is referred to as the projection of $\mathbf{b}$ onto S. This idea is central to the least squares and approximation applications in the following sections of the chapter. The primary examples of orthogonal subspaces and orthogonal complements are the two pairs of orthogonal subspaces associated with each $m \times n$ matrix.

Important Concepts

1. **Orthogonal subspaces.** Two subspaces X and Y of R^n are said to be *orthogonal* if $\mathbf{x}^T\mathbf{y} = 0$ for every $\mathbf{x} \in X$ and every $\mathbf{y} \in Y$. If X and Y are orthogonal, we write $X \perp Y$.

2. **Orthogonal complements.** Let Y be a subspace of R^n. The set of all vectors in R^n that are orthogonal to every vector in Y will be denoted $Y^\perp$. Thus
$$Y^\perp = \left\{\mathbf{x} \in R^n \mid \mathbf{x}^T\mathbf{y} = 0 \quad \text{for every} \quad \mathbf{y} \in Y\right\}$$
The set $Y^\perp$ is called the *orthogonal complement* of Y. It is easy to see that if Y is a subspace of R^n, then $Y^\perp$ will also be a subspace of R^n.

3. **Fundamental subspaces.** Let A be an $m \times n$ matrix. There are four *fundamental subspaces* associated with A. The first two, the column space of A, denoted $R(A)$, and the nullspace of A^T, denoted $N(A^T)$, are both of subspaces of R^m. The other two, the column space of A^T and the nullspace of A are both subspaces of R^n.

4. **Direct sum.** If U and V are subspaces of a vector space W and each $\mathbf{w} \in W$ can be written uniquely as a sum $\mathbf{u} + \mathbf{v}$, where $\mathbf{u} \in U$ and $\mathbf{v} \in V$, then we say that W is a *direct sum* of U and V, and we write $W = U \oplus V$.

Important Theorems

▶ THEOREM 5.2.1 **(Fundamental Subspaces Theorem)** *If A is an $m \times n$ matrix, then $N(A) = R(A^T)^\perp$ and $N(A^T) = R(A)^\perp$.*

▶ THEOREM 5.2.2. *If S is a subspace of R^n, then $\dim S + \dim S^\perp = n$. Furthermore, if $\{\mathbf{x}_1, \ldots, \mathbf{x}_r\}$ is a basis for S and $\{\mathbf{x}_{r+1}, \ldots, \mathbf{x}_n\}$ is a basis for $S^\perp$, then $\{\mathbf{x}_1, \ldots, \mathbf{x}_r, \mathbf{x}_{r+1}, \ldots, \mathbf{x}_n\}$ is a basis for R^n.*

▶ THEOREM 5.2.3. *If S is a subspace of R^n, then*
$$R^n = S \oplus S^\perp$$

▶ THEOREM 5.2.4. *If S is a subspace of R^n, then $(S^\perp)^\perp = S$.*

▶ COROLLARY 5.2.5. *If A is an $m \times n$ matrix and $\mathbf{b} \in R^m$, then either there is a vector $\mathbf{x} \in R^n$ such that $A\mathbf{x} = \mathbf{b}$ or there is a vector $\mathbf{y} \in R^m$ such that $A^T\mathbf{y} = \mathbf{0}$ and $\mathbf{y}^T\mathbf{b} \neq 0$.*

Exercises: Solutions and Hints

1. (b) **Solution.** The reduced row echelon form of A is

$$U = \begin{pmatrix} 1 & 0 & -2 \\ 0 & 1 & 1 \end{pmatrix}$$

If we transpose U, the vectors $(1,\ 0,\ -2)^T$, $(0,\ 1,\ 1)^T$ will form a basis for the column space of U^T. However, we could have used column operations to reduce A^T to U^T. So A^T and U^T have the same column space. Thus $\{(1,\ 0,\ -2)^T,\ (0,\ 1,\ 1)^T\}$ is a basis for the column space of A^T. To find the nullspace we must solve $A\mathbf{x} = \mathbf{0}$ which is equivalent to $U\mathbf{x} = \mathbf{0}$. The augmented matrix for the reduced system is

$$\begin{pmatrix} 1 & 0 & -2 & 0 \\ 0 & 1 & 1 & 0 \end{pmatrix}$$

There is one free variable $\mathbf{x}_3$. Solving for $\mathbf{x}_1$ and $\mathbf{x}_2$ in terms of x_3, we get

$$\mathbf{x}_1 = 2\mathbf{x}_3 \qquad \text{and} \qquad \mathbf{x}_2 = -\mathbf{x}_3$$

The solution set consists of all vectors of the form

$$\begin{pmatrix} 2\mathbf{x}_3 \\ -\mathbf{x}_3 \\ \mathbf{x}_3 \end{pmatrix} = x_3 \begin{pmatrix} 2 \\ -1 \\ 1 \end{pmatrix}$$

Thus $\{(2, -1, 1)^T\}$ is a basis for $N(A)$.

To find bases for $R(A)$ and $N(A^T)$ we compute the reduced row echelon form of A^T,

$$V = \begin{pmatrix} 1 & 0 \\ 0 & 1 \\ 0 & 0 \end{pmatrix}$$

V^T and A have the same column space which is spanned by

$$\mathbf{e}_1 = \begin{pmatrix} 1 \\ 0 \end{pmatrix} \qquad \text{and} \qquad \mathbf{e}_2 = \begin{pmatrix} 0 \\ 1 \end{pmatrix}$$

so $R(A) = R^2$. To find the $N(A^T)$, we must solve $A^T\mathbf{x} = \mathbf{0}$. This system is equivalent to the reduced system $V\mathbf{x} = \mathbf{0}$. The augmented matrix for the reduced system is

$$\left(\begin{array}{cc|c} 1 & 0 & 0 \\ 0 & 1 & 0 \\ 0 & 0 & 0 \end{array} \right)$$

The only solution is the zero vector. Thus $N(A^T) = \{(0, 0)^T\}$.

2. (a) **Hint.** A vector $\mathbf{y}$ is in $S^\perp$ if and only if $\mathbf{y}^T\mathbf{x} = 0$.

3. (a) **Hint.** A vector $\mathbf{z}$ will be in $S^\perp$ if and only if $\mathbf{z}$ is orthogonal to both $\mathbf{x}$ and $\mathbf{y}$.

6. **Hint.** If $(3, 1, 2)$ is in the row space of A, then $(3, 1, 2)^T$ is in the column space of A^T.

7. **Hint.** What do we know about the intersection of $N(A^T)$ and $R(A)$?

10. **Hint.** If $A\mathbf{x} = \mathbf{b}$ has no solution then $\mathbf{b} \notin R(A)$. But $R(A) = N(A^T)^\perp$, so $\mathbf{b} \notin N(A^T)^\perp$.

11. **Hint.** The argument here is similar to that used in the previous exercise.

12. **Hint.** Make use of Theorem 5.2.3.

13. (a) **Hint.** $A\mathbf{x} = x_1\mathbf{a}_1 + x_1\mathbf{a}_1 + \cdots + x_n\mathbf{a}_n$.
    (b) **Hint.** You need to show that if $\mathbf{x} \in N(A)$ then $\mathbf{x} \in N(A^T A)$ and also if $\mathbf{x} \in N(A^T A)$ then $\mathbf{x} \in N(A)$. One of these statements should be very easy to show and the other is easy to show if you use the result from part (a).

15. **Hint.** If $\mathbf{x} \in U \cap V$, then we can write

$$\begin{array}{ll} \mathbf{x} = \mathbf{0} + \mathbf{x} & (\mathbf{0} \in U, \quad \mathbf{x} \in V) \\ \mathbf{x} = \mathbf{x} + \mathbf{0} & (\mathbf{x} \in U, \quad \mathbf{0} \in V) \end{array}$$

16. **Hint.** You need to show that $A\mathbf{x}_1, A\mathbf{x}_2, \ldots, A\mathbf{x}_r$ are all in $R(A)$ and that they are linearly independent. To do this it is useful to note that

$$R(A^T) \cap N(A) = \{\mathbf{0}\}$$

17. (b) **Hint.** If $\mathbf{z}$ is in $N(A)$ then

$$\mathbf{0} = A\mathbf{z} = \mathbf{x}\mathbf{y}^T\mathbf{z} + \mathbf{y}\mathbf{x}^T\mathbf{z} = (\mathbf{y}^T\mathbf{z})\mathbf{x} + (\mathbf{x}^T\mathbf{z})\mathbf{y}$$

3 | LEAST SQUARES PROBLEMS

Overview

A standard technique in mathematical and statistical modeling is to find a least fit to a set of data points. The least squares problem can be formulated as an overdetermined linear system $A\mathbf{x} = \mathbf{b}$. A vector $\mathbf{x}$ is a least squares solution to the problem if it minimizes the distance between $A\mathbf{x}$ and $\mathbf{b}$. In this section we present a method for solving the least squares problem and characterize when the problem has a unique solution.

Important Concepts

1. **Least squares problem.** Given a system of equations $A\mathbf{x} = \mathbf{b}$, where A is an $m \times n$ matrix with $m > n$ and $\mathbf{b} \in R^m$, then for each $\mathbf{x} \in R^n$ we can form a *residual*

$$r(\mathbf{x}) = \mathbf{b} - A\mathbf{x}$$

The distance between $\mathbf{b}$ and $A\mathbf{x}$ is given by

$$\|\mathbf{b} - A\mathbf{x}\| = \|r(\mathbf{x})\|$$

The *least squares problem* is to find a vector $\mathbf{x} \in R^n$ for which $\|r(\mathbf{x})\|$ will be a minimum. Minimizing $\|r(\mathbf{x})\|$ is equivalent to minimizing $\|r(\mathbf{x})\|^2$. A vector $\hat{\mathbf{x}}$ that accomplishes this is said to be a *least squares solution* to the system $A\mathbf{x} = \mathbf{b}$.

2. **Projection matrix.** If A is a $m \times n$ of rank n, then the matrix

$$P = A(A^T A)^{-1} A^T$$

is called the *projection matrix* that projects vectors in R^m onto the column space of A. Thus if $\mathbf{b}$ is any vector in $R(A)$ and $\mathbf{p} = P\mathbf{b}$, then $\mathbf{p}$ is the vector in the column space of A that is closest to $\mathbf{b}$.

Important Theorems

▶ THEOREM 5.3.1. *Let S be a subspace of R^m. For each $\mathbf{b} \in R^m$ there is a unique element $\mathbf{p}$ of S that is closest to $\mathbf{b}$, that is,*

$$\|\mathbf{b} - \mathbf{y}\| > \|\mathbf{b} - \mathbf{p}\|$$

for any $\mathbf{y} \neq \mathbf{p}$ in S. Furthermore, a given vector $\mathbf{p}$ in S will be closest to a given vector $\mathbf{b} \in R^m$ if and only if $\mathbf{b} - \mathbf{p} \in S^\perp$.

▶ THEOREM 5.3.2. *If A is an $m \times n$ matrix of rank n, the normal equations*

$$A^T A\mathbf{x} = A^T \mathbf{b}$$

have a unique solution

$$\hat{\mathbf{x}} = (A^T A)^{-1} A^T \mathbf{b}$$

and $\hat{\mathbf{x}}$ is the unique least squares solution to the system $A\mathbf{x} = \mathbf{b}$.

Exercises: Solutions and Hints

1. **Solution.**

(b) $A^T A = \begin{bmatrix} 6 & -1 \\ -1 & 6 \end{bmatrix}$ and $A^T \mathbf{b} = \begin{bmatrix} 20 \\ -25 \end{bmatrix}$.

The solution to the normal equations $A^T A \mathbf{x} = A^T \mathbf{b}$ is

$$\mathbf{x} = \begin{pmatrix} \frac{19}{7} \\ -\frac{26}{7} \end{pmatrix}$$

7. **Solution.** To find the best fit by a linear function we must find the least squares solution to the linear system

$$\begin{pmatrix} 1 & x_1 \\ 1 & x_2 \\ \vdots & \vdots \\ 1 & x_n \end{pmatrix} \begin{pmatrix} c_0 \\ c_1 \end{pmatrix} = \begin{pmatrix} y_1 \\ y_2 \\ \vdots \\ y_n \end{pmatrix}$$

If we form the normal equations the augmented matrix for the system will be

$$\begin{pmatrix} n & \sum_{i=1}^{n} x_i & \bigg| & \sum_{i=1}^{n} y_i \\ \sum_{i=1}^{n} x_i & \sum_{i=1}^{n} x_i^2 & \bigg| & \sum_{i=1}^{n} x_i y_i \end{pmatrix}$$

If $\bar{x} = 0$ then

$$\sum_{i=1}^{n} x_i = n\bar{x} = 0$$

and hence the coefficient matrix for the system is diagonal. The solution is easily obtained.

$$c_0 = \frac{\sum_{i=1}^{n} y_i}{n} = \bar{y}$$

and

$$c_1 = \frac{\sum_{i=1}^{n} x_i y_i}{\sum_{i=1}^{n} x_i^2} = \frac{\mathbf{x}^T \mathbf{y}}{\mathbf{x}^T \mathbf{x}}$$

9. (a) **Hint.** If $\mathbf{b} \in R(A)$ then $\mathbf{b} = A\mathbf{x}$ for some $\mathbf{x} \in R^n$.

   (b) **Hint.** $R(A)^{\perp} = N(A^T)$.

   (c) **Solution.** The following figures give a geometric illustration of parts (a) and (b). In the first figure $\mathbf{b}$ lies in the plane corresponding to $R(A)$. Since it is already in the plane, projecting it onto the plane will have no effect. In the second figure $\mathbf{b}$ lies on the line through the origin that is normal to the plane. When it is projected onto the plane it projects right down to the origin.

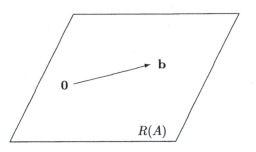

If $\mathbf{b} \in R(A)$, then $P\mathbf{b} = \mathbf{b}$.

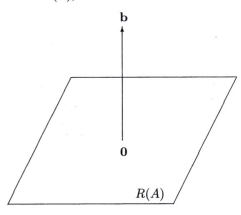

If $\mathbf{b} \in R(A)^{\perp}$, then $P\mathbf{b} = \mathbf{0}$.

10. **Hint.** If the system were consistent what would that tell us about $\mathbf{b}$ in relation to one of the fundamental subspaces?

13. **Solution.** If $\hat{\mathbf{x}}$ is a solution to the least squares problem, then $\hat{\mathbf{x}}$ is a solution to the normal equations

$$A^T A\mathbf{x} = A^T \mathbf{b}$$

It follows that a vector $\mathbf{y} \in R^n$ will be a solution if and only if

$$\mathbf{y} = \hat{\mathbf{x}} + \mathbf{z}$$

for some $\mathbf{z} \in N(A^T A)$. (See Exercise 20, Chapter 3, Section 6). Since

$$N(A^T A) = N(A)$$

we conclude that $\mathbf{y}$ is a least squares solution if and only if

$$\mathbf{y} = \hat{\mathbf{x}} + \mathbf{z}$$

for some $\mathbf{z} \in N(A)$.

---

4 | SECTION INNER PRODUCT SPACES

Overview

Scalar products are useful not only in R^n but in a wide variety of contexts. To generalize scalar products to other vector spaces we introduce the concept of an inner product. An inner product in a vector space V works exactly like the scalar product works in R^n. In particular, the inner product can be used to define lengths of vectors, distance between vectors, angles between vectors, orthogonality, and vector projections.

Important Concepts

1. **Inner product and inner product space.** An *inner product* on a vector space V is an operation on V that assigns to each pair of vectors $\mathbf{x}$ and $\mathbf{y}$ in V a real number $\langle \mathbf{x}, \mathbf{y} \rangle$ satisfying the following conditions:
   (i) $\langle \mathbf{x}, \mathbf{x} \rangle \geq 0$ with equality if and only if $\mathbf{x} = \mathbf{0}$.
   (ii) $\langle \mathbf{x}, \mathbf{y} \rangle = \langle \mathbf{y}, \mathbf{x} \rangle$ for all $\mathbf{x}$ and $\mathbf{y}$ in V.
   (iii) $\langle \alpha \mathbf{x} + \beta \mathbf{y}, \mathbf{z} \rangle = \alpha \langle \mathbf{x}, \mathbf{z} \rangle + \beta \langle \mathbf{y}, \mathbf{z} \rangle$ for all $\mathbf{x}, \mathbf{y}, \mathbf{z}$ in V and all scalars α and β.
   A vector space V with an inner product is called an *inner product space*.

2. **Inner products for R^n.** The standard inner product for R^n is the scalar product
$$\langle \mathbf{x}, \mathbf{y} \rangle = \mathbf{x}^T \mathbf{y}$$
Given a vector $\mathbf{w}$ with positive entries, we could also define an inner product on R^n by
$$\langle \mathbf{x}, \mathbf{y} \rangle = \sum_{i=1}^{n} x_i y_i w_i$$
The entries w_i are referred to as *weights*.

3. **Inner product for $R^{m \times n}$.** Given A and B in $R^{m \times n}$, one can define an inner product by
$$\langle A, B \rangle = \sum_{i=1}^{m} \sum_{j=1}^{n} a_{ij} b_{ij}$$

4. **Inner products for $C[a, b]$.** The standard inner product for $C[a, b]$ is defined by
$$\langle f, g \rangle = \int_a^b f(x) g(x) \, dx$$
If $w(x)$ is a positive continuous function on $[a, b]$, then
$$\langle f, g \rangle = \int_a^b f(x) g(x) w(x) \, dx$$

also defines an inner product on $C[a, b]$. The function $w(x)$ is called a *weight function*. In particular for $C[-\pi, \pi]$ it is useful in many applications to choose the weight function to be the constant function $w(x) = \dfrac{1}{\pi}$.

5. **Discrete inner product for P_n.** Given any n distinct real numbers, $x_1, x_2, \ldots, x_n$, we can define an inner product on P_n by

$$\langle p, q \rangle = \sum_{i=1}^{n} p(x_i) q(x_i)$$

6. **Norm of a vector.** If $\mathbf{v}$ is a vector in an inner product space V, the *length* or *norm* of $\mathbf{v}$ is given by
$$\|\mathbf{v}\| = \sqrt{\langle \mathbf{v}, \mathbf{v} \rangle}$$

7. **Orthogonal vectors.** Two vectors $\mathbf{u}$ and $\mathbf{v}$ in an inner product space are said to be *orthogonal* if $\langle \mathbf{u}, \mathbf{v} \rangle = 0$.

8. **Frobenius norm of a matrix.** For the vector space $R^{m \times n}$ the norm derived from the inner product is called the *Frobenius norm* and is denoted by $\| \cdot \|_F$. Thus, if $A \in R^{m \times n}$, then

$$\|A\|_F = (\langle A, A \rangle)^{1/2} = \left(\sum_{i=1}^{m} \sum_{j=1}^{n} a_{ij}^2 \right)^{1/2}$$

9. **Vector and scalar projections.** If $\mathbf{u}$ and $\mathbf{v}$ are vectors in an inner product space V and $\mathbf{v} \neq \mathbf{0}$, then the *scalar projection* of $\mathbf{u}$ onto $\mathbf{v}$ is given by

$$\alpha = \frac{\langle \mathbf{u}, \mathbf{v} \rangle}{\|\mathbf{v}\|}$$

and the *vector projection* of $\mathbf{u}$ onto $\mathbf{v}$ is given by

$$\mathbf{p} = \alpha \left(\frac{1}{\|\mathbf{v}\|} \mathbf{v} \right) = \frac{\langle \mathbf{u}, \mathbf{v} \rangle}{\langle \mathbf{v}, \mathbf{v} \rangle} \mathbf{v}$$

10. **Angle between vectors.** If $\mathbf{u}$ and $\mathbf{v}$ are vectors in an inner product space, then the angle between the vectors is the angle θ in the interval $[0, \pi]$ that satisfies

$$\cos \theta = \frac{\langle \mathbf{u}, \mathbf{v} \rangle}{\|\mathbf{u}\| \|\mathbf{v}\|}$$

11. **General norms.** A vector space V is said to be a *normed linear space* if to each vector $\mathbf{v} \in V$ there is associated a real number $\|\mathbf{v}\|$, called the *norm* of $\mathbf{v}$, satisfying:
    (i) $\|\mathbf{v}\| \geq 0$ with equality if and only if $\mathbf{v} = \mathbf{0}$.
    (ii) $\|\alpha \mathbf{v}\| = |\alpha| \, \|\mathbf{v}\|$ for any scalar α.
    (iii) $\|\mathbf{v} + \mathbf{w}\| \leq \|\mathbf{v}\| + \|\mathbf{w}\|$ for all $\mathbf{v}, \mathbf{w} \in V$.
    The third condition is called the *triangle inequality*

12. **2-norm.** The *2-norm* on R^n is the norm derived from the standard inner product. Thus

$$\|\mathbf{x}\|_2 = \sqrt{\mathbf{x}^T \mathbf{x}} = (x_1^2 + x_2^2 + \cdots + x_n^2)^{\frac{1}{2}}$$

13. 1-norm. The 1-*norm* for R^n is defined by

$$\|\mathbf{x}\|_1 = \sum_{i=1}^{n} |x_i|$$

14. ∞-norm. The *uniform norm* or *infinity norm* on R^n is defined by

$$\|\mathbf{x}\|_\infty = \max_{1 \le i \le n} |x_i|$$

15. Distance between vectors. Let $\mathbf{x}$ and $\mathbf{y}$ be vectors in a normed linear space. The *distance between* $\mathbf{x}$ *and* $\mathbf{y}$ is defined to be the number $\|\mathbf{y} - \mathbf{x}\|$.

Important Theorems

▶ THEOREM 5.4.1 **(The Pythagorean Law)** *If* $\mathbf{u}$ *and* $\mathbf{v}$ *are orthogonal vectors in an inner product space* V, *then*

$$\|\mathbf{u} + \mathbf{v}\|^2 = \|\mathbf{u}\|^2 + \|\mathbf{v}\|^2$$

▶ THEOREM 5.4.2 **(The Cauchy–Schwarz Inequality)** *If* $\mathbf{u}$ *and* $\mathbf{v}$ *are any two vectors in an inner product space* V, *then*

(1) $$|\langle \mathbf{u}, \mathbf{v} \rangle| \le \|\mathbf{u}\|\,\|\mathbf{v}\|$$

Equality holds if and only if $\mathbf{u}$ *and* $\mathbf{v}$ *are linearly dependent.*

▶ THEOREM 5.4.3. *If* V *is an inner product space, then the equation*

$$\|\mathbf{v}\| = \sqrt{\langle \mathbf{v}, \mathbf{v} \rangle} \qquad \text{for all} \quad \mathbf{v} \in V$$

defines a norm on V.

Exercises: Solutions and Hints

2. (b) Solution.

$$\mathbf{p} = \frac{\mathbf{x}^T \mathbf{y}}{\mathbf{y}^T \mathbf{y}} \mathbf{y} = \frac{12}{72} \mathbf{y} = \left(\frac{4}{3}, \frac{1}{3}, \frac{1}{3}, 0 \right)^T$$

5. Solution. We must show that the three conditions from the definition of an inner product are satisfied.

(i)

$$\langle A, A \rangle = \sum_{i=1}^{m} \sum_{j=1}^{n} a_{ij}^2 \ge 0$$

and $\langle A, A \rangle = 0$ if and only if each $a_{ij} = 0$.

(ii) $$\langle A, B \rangle = \sum_{i=1}^{m} \sum_{j=1}^{n} a_{ij} b_{ij} = \sum_{i=1}^{m} \sum_{j=1}^{n} b_{ij} a_{ij} = \langle B, A \rangle$$

(iii)

$$\langle \alpha A + \beta B, C \rangle = \sum_{i=1}^{m} \sum_{j=1}^{n} (\alpha a_{ij} + \beta b_{ij}) c_{ij}$$

$$= \alpha \sum_{i=1}^{m} \sum_{j=1}^{n} a_{ij} c_{ij} + \beta \sum_{i=1}^{m} \sum_{j=1}^{n} b_{ij} c_{ij}$$

$$= \alpha \langle A, C \rangle + \beta \langle B, C \rangle$$

9. Solution. The vectors $\cos mx$ and $\sin nx$ are orthogonal since

$$
\begin{aligned}
\langle \cos mx, \sin nx \rangle &= \frac{1}{\pi} \int_{-\pi}^{\pi} \cos mx \sin nx \, dx \\
&= \frac{1}{2\pi} \int_{-\pi}^{\pi} [\sin(n+m)x + \sin(n-m)x] \, dx \\
&= 0
\end{aligned}
$$

They are unit vectors since

$$
\begin{aligned}
\langle \cos mx, \cos mx \rangle &= \frac{1}{\pi} \int_{-\pi}^{\pi} \cos^2 mx \, dx \\
&= \frac{1}{2\pi} \int_{-\pi}^{\pi} [1 + \cos 2mx] \, dx \\
&= 1
\end{aligned}
$$

$$
\begin{aligned}
\langle \sin nx, \sin nx \rangle &= \frac{1}{\pi} \int_{-\pi}^{\pi} \sin nx \sin nx \, dx \\
&= \frac{1}{2\pi} \int_{-\pi}^{\pi} (1 - \cos 2nx) \, dx \\
&= 1
\end{aligned}
$$

Since the $\cos mx$ and $\sin nx$ are orthogonal, the distance between the vectors can be determined using the Pythagorean law.

$$\| \cos mx - \sin nx \| = (\| \cos mx \|^2 + \| \sin nx \|^2)^{\frac{1}{2}} = \sqrt{2}$$

14. Solution.
  (i) $\| \mathbf{x} \|_\infty = \max_{1 \leq i \leq n} |x_i| \geq 0$. If $\max_{1 \leq i \leq n} |x_i| = 0$ then all of the x_i's must be zero and hence $\mathbf{x} = \mathbf{0}$.
  (ii) $\| \alpha \mathbf{x} \|_\infty = \max_{1 \leq i \leq n} |\alpha x_i| = |\alpha| \max_{1 \leq i \leq n} |x_i| = |\alpha| \, \| \mathbf{x} \|_\infty$
  (iii) $\| \mathbf{x} + \mathbf{y} \|_\infty = \max |x_i + y_i| \leq \max |x_i| + \max |y_i| = \| \mathbf{x} \|_\infty + \| \mathbf{y} \|_\infty$

17. Hint. If $\mathbf{x}$ is orthogonal to $\mathbf{y}$, then it is also orthogonal to $-\mathbf{y}$. Note that $\| \mathbf{x} - \mathbf{y} \| = \| \mathbf{x} + (-\mathbf{y}) \|$.

22. Solution. $\| - \mathbf{v} \| = \| (-1)\mathbf{v} \| = |-1| \, \| \mathbf{v} \| = \| \mathbf{v} \|$

24. Solution.

$$
\begin{aligned}
\| \mathbf{u} + \mathbf{v} \|^2 &= \langle \mathbf{u} + \mathbf{v}, \mathbf{u} + \mathbf{v} \rangle \\
&= \langle \mathbf{u}, \mathbf{u} \rangle + \langle \mathbf{u}, \mathbf{v} \rangle + \langle \mathbf{v}, \mathbf{u} \rangle + \langle \mathbf{v}, \mathbf{v} \rangle \\
&= \| \mathbf{u} \|^2 + 2 \langle \mathbf{u}, \mathbf{v} \rangle + \| \mathbf{v} \|^2
\end{aligned}
$$

Similarly

$$\|\mathbf{u} - \mathbf{v}\|^2 = \|\mathbf{u}\|^2 - 2\langle \mathbf{u}, \mathbf{v}\rangle + \|\mathbf{v}\|^2$$

Adding the two equations,

$$\|\mathbf{u} + \mathbf{v}\|^2 = \|\mathbf{u}\|^2 + 2\langle \mathbf{u}, \mathbf{v}\rangle + \|\mathbf{v}\|^2$$
$$\|\mathbf{u} - \mathbf{v}\|^2 = \|\mathbf{u}\|^2 - 2\langle \mathbf{u}, \mathbf{v}\rangle + \|\mathbf{v}\|^2$$

we see that

$$\|\mathbf{u} + \mathbf{v}\|^2 + \|\mathbf{u} - \mathbf{v}\|^2 = 2\|\mathbf{u}\|^2 + 2\|\mathbf{v}\|^2$$

If the vectors $\mathbf{u}$ and $\mathbf{v}$ are used to form a parallelogram in the plane, then the diagonals will be $\mathbf{u} + \mathbf{v}$ and $\mathbf{u} - \mathbf{v}$. The equation shows that the sum of the squares of the lengths of the diagonals equals the sum of the squares of the lengths of the four sides. This result is referred as the *parallelogram rule*.

26. (a) **Hint.** If

$$|f(a)| + |f(b)| = 0$$

does this imply that f must be the zero function?

(b) **Hint.** This is how the 1-norm is defined for function spaces. It is the continuous version of the discrete 1-norm we use for R^n. Show that the 3 conditions required in the definition of a norm are satisfied.

(c) **Hint.** This is how the infinity norm is defined for function spaces. It is the continuous version of the discrete infinity norm we use for R^n. Show that the 3 conditions required in the definition of a norm are satisfied.

28. **Solution.** Each norm produces a different unit "circle".

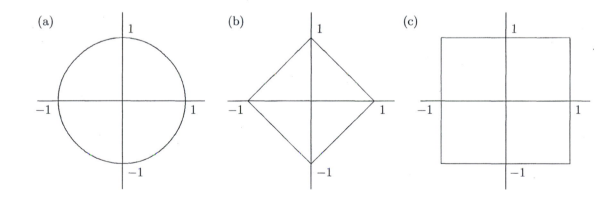

5 | ORTHONORMAL SETS

Overview

The standard basis vectors for R^n are mutually orthogonal unit vectors. Such a basis is called an orthonormal basis. There are many advantages to using an orthonormal basis to represent a vector space or a subspace. In particular it turns out to be much easier to represent vectors as linear combinations of the bases vectors if you have an orthonormal basis for the vector space. It is also much easier to project onto a subspace if you have an orthonormal basis for the subspace. Furthermore, matrices whose column vectors form orthonormal bases for R^n have many useful properties. Such matrices are called orthogonal matrices. These matrices are an important tool in the computational solution of many mathematical problems.

Important Concepts

1. **Orthogonal set of vectors.** Let $\mathbf{v}_1, \mathbf{v}_2, \ldots, \mathbf{v}_n$ be vectors in an inner product space V. If $\langle \mathbf{v}_i, \mathbf{v}_j \rangle = 0$ whenever $i \neq j$, then $\{\mathbf{v}_1, \mathbf{v}_2, \ldots, \mathbf{v}_n\}$ is said to be an *orthogonal set* of vectors.

2. **Orthonormal set of vectors.** An *orthonormal set* of vectors is an orthogonal set of unit vectors. The set $\{\mathbf{u}_1, \mathbf{u}_2, \ldots, \mathbf{u}_n\}$ will be orthonormal if and only if
$$\langle \mathbf{u}_i, \mathbf{u}_j \rangle = \delta_{ij}$$
where
$$\delta_{ij} = \begin{cases} 1 & \text{if } i = j \\ 0 & \text{if } i \neq j \end{cases}$$

3. **Orthogonal matrices.** An $n \times n$ matrix Q is said to be an *orthogonal matrix* if the column vectors of Q form an orthonormal set in R^n.

4. **Permutation matrices.** A *permutation matrix* is a matrix formed from the identity matrix by reordering its columns.

5. **Projection onto a subspace.** If S is a subspace of an inner product space V, then for each $\mathbf{v}$ in V there is a unique vector $\mathbf{p}$ in S that is closest to $\mathbf{v}$. The vector $\mathbf{p}$ is referred to as the *vector projection* of $\mathbf{v}$ onto S.

6. **Projection matrix.** For each subspace S of R^m, there exists a unique matrix P such that for all $\mathbf{b} \in R^m$, the vector $\mathbf{p} = P\mathbf{b}$ is the projection of $\mathbf{b}$ onto S. The matrix P with this property is referred to as the *projection matrix* corresponding to the subspace S.

Important Theorems and Results

▶ THEOREM 5.5.1. *If $\{\mathbf{v}_1, \mathbf{v}_2, \ldots, \mathbf{v}_n\}$ is an orthogonal set of nonzero vectors in an inner product space V, then $\mathbf{v}_1, \mathbf{v}_2, \ldots, \mathbf{v}_n$ are linearly independent.*

▶ THEOREM 5.5.2. *Let $\{\mathbf{u}_1, \mathbf{u}_2, \ldots, \mathbf{u}_n\}$ be an orthonormal basis for an inner product space V. If $\mathbf{v} = \sum_{i=1}^{n} c_i \mathbf{u}_i$, then $c_i = \langle \mathbf{u}_i, \mathbf{v} \rangle$.*

▶ **COROLLARY 5.5.3.** *Let* $\{\mathbf{u}_1, \mathbf{u}_2, \ldots, \mathbf{u}_n\}$ *be an orthonormal basis for an inner product space* V. *If* $\mathbf{u} = \sum_{i=1}^{n} a_i \mathbf{u}_i$ *and* $\mathbf{v} = \sum_{i=1}^{n} b_i \mathbf{u}_i$, *then*

$$\langle \mathbf{u}, \mathbf{v} \rangle = \sum_{i=1}^{n} a_i b_i$$

▶ **COROLLARY 5.5.4 (Parseval's Formula)** *If* $\{\mathbf{u}_1, \ldots, \mathbf{u}_n\}$ *is an orthonormal basis for an inner product space* V *and* $\mathbf{v} = \sum_{i=1}^{n} c_i \mathbf{u}_i$, *then*

$$\|\mathbf{v}\|^2 = \sum_{i=1}^{n} c_i^2$$

▶ **THEOREM 5.5.5.** *An* $n \times n$ *matrix* Q *is orthogonal if and only if* $Q^T Q = I$.

▶ **Properties of Orthogonal Matrices.**
If Q is an $n \times n$ orthogonal matrix, then:

(a) The column vectors of Q form an orthonormal basis for R^n.
(b) $Q^T Q = I$
(c) $Q^T = Q^{-1}$
(d) $\langle Q\mathbf{x}, Q\mathbf{y} \rangle = \langle \mathbf{x}, \mathbf{y} \rangle$
(e) $\|Q\mathbf{x}\|_2 = \|\mathbf{x}\|_2$
(f) For any vectors $\mathbf{x}$ and $\mathbf{y}$ in R^n the angle between $Q\mathbf{x}$ and $Q\mathbf{y}$ is equal to the angle between $\mathbf{x}$ and $\mathbf{y}$.

▶ **THEOREM 5.5.6.** *If the column vectors of* A *form an orthonormal set of vectors in* R^m, *then* $A^T A = I$ *and the solution to the least squares problem is*

$$\hat{\mathbf{x}} = A^T \mathbf{b}$$

▶ **THEOREM 5.5.7.** *Let* S *be a subspace of an inner product space* V *and let* $\mathbf{x} \in V$. *Let* $\{\mathbf{x}_1, \mathbf{x}_2, \ldots, \mathbf{x}_n\}$ *be an orthonormal basis for* S. *If*

(1) $$\mathbf{p} = \sum_{i=1}^{n} c_i \mathbf{x}_i$$

where
(2) $$c_i = \langle \mathbf{x}, \mathbf{x}_i \rangle \qquad \textit{for each } i$$
then $\mathbf{p} - \mathbf{x} \in S^\perp$.

▶ **THEOREM 5.5.8.** *Under the hypothesis of Theorem 5.5.7,* $\mathbf{p}$ *is the element of* S *that is closest to* $\mathbf{x}$, *that is,*

$$\|\mathbf{y} - \mathbf{x}\| > \|\mathbf{p} - \mathbf{x}\|$$

for any $\mathbf{y} \neq \mathbf{p}$ *in* S.

▶ COROLLARY 5.5.9. *Let S be a nonzero subspace of R^m and let $\mathbf{b} \in R^m$. If $\{\mathbf{u}_1, \mathbf{u}_2, \ldots, \mathbf{u}_k\}$ is an orthonormal basis for S and $U = (\mathbf{u}_1, \mathbf{u}_2, \ldots, \mathbf{u}_k)$, then the projection $\mathbf{p}$ of $\mathbf{b}$ onto S is given by*

$$\mathbf{p} = UU^T\mathbf{b}$$

Exercises: Solutions and Hints

2. (a) **Hint.** You need to show that the vectors are mutually orthogonal, i.e.,

$$\mathbf{u}_1^T\mathbf{u}_2 = \mathbf{u}_1^T\mathbf{u}_3 = \mathbf{u}_2^T\mathbf{u}_3 = 0$$

and also that the vectors are unit vectors, i.e.,

$$\mathbf{u}_1^T\mathbf{u}_1 = \mathbf{u}_2^T\mathbf{u}_2 = \mathbf{u}_3^T\mathbf{u}_3 = 1$$

4. (b) **Hint.** Make use of Theorem 5.5.2.

5. **Hint.** Make use of Theorem 5.5.2 and Parseval's formula.

7. See the hint for Exercise 5.

8. **Solution.** Since $\{\sin x, \cos x\}$ is an orthonormal set it follows that

$$\langle f, g \rangle = 3 \cdot 1 + 2 \cdot (-1) = 1$$

9. (a) **Solution.** $\sin^4 x = \left(\dfrac{1 - \cos 2x}{2} \right)^2$

$$= \frac{1}{4}\cos^2 2x - \frac{1}{2}\cos 2x + \frac{1}{4}$$

$$= \frac{1}{4}\left(\frac{1 + \cos 4x}{2} \right) - \frac{1}{2}\cos 2x + \frac{1}{4}$$

$$= \frac{1}{8}\cos 4x - \frac{1}{2}\cos 2x + \frac{3\sqrt{2}}{8}\frac{1}{\sqrt{2}}$$

(b) (ii) **Solution.** Using Theorem 5.5.2, we have

$$\int_{-\pi}^{\pi} \sin^4 x \cos 2x \, dx = \pi\langle \sin^4 x, \cos 2x \rangle = \pi \cdot \left(-\frac{1}{2} \right) = -\frac{\pi}{2}$$

14. **Hint.** H is symmetric since

$$H^T = (I - 2\mathbf{u}\mathbf{u}^T)^T = I^T - 2(\mathbf{u}^T)^T\mathbf{u}^T = I - 2\mathbf{u}\mathbf{u}^T = H$$

Since H is symmetric it follows that $H^T H = H^2$. So, to show that H is orthogonal you need to show that $H^2 = I$.

15. **Hint.** Since $\det(Q^T) = \det(Q)$, it follows that

$$[\det(Q)]^2 = \det(Q^T)\det(Q)$$

16. (b) **Solution.** The product of two permutation matrices will be a permutation matrix. To see this let P_1 and P_2 be permutation matrices. The columns of P_1 are the same as the columns of I, but in a different order. Postmultiplication of P_1 by P_2 reorders the columns of P_1. Thus

$P_1 P_2$ is a matrix formed by reordering the columns of I and hence is a permutation matrix.

18. Hint. If P is a symmetric permutation matrix then P is orthogonal and $P^{-1} = P^T = P$.

24. (a) **Solution.** Let U_1 be a matrix whose columns form an orthonormal basis for $R(A)$ and let U_2 be a matrix whose columns form an orthonormal basis for $N(A^T)$. If we set $U = (U_1, U_2)$, then since $R(A)$ and $N(A^T)$ are orthogonal complements in R^n, it follows that U is an orthogonal matrix. The unique projection matrix P onto $R(A)$ is given $P = U_1 U_1^T$ and the projection matrix onto $N(A^T)$ is given by $U_2 U_2^T$. Since U is orthogonal it follows that

$$I = UU^T = U_1 U_1^T + U_2 U_2^T = P + U_2 U_2^T$$

Thus the projection matrix onto $N(A^T)$ is given by

$$U_2 U_2^T = I - P$$

25. (a) **Hint.** If U is a matrix whose columns form an orthonormal basis for S, then the projection matrix P corresponding to S is given by $P = UU^T$.

26. Hint. Show that $A^T A$ is a diagonal matrix and that its diagonal entries are $\mathbf{a}_1^T \mathbf{a}_1, \mathbf{a}_2^T \mathbf{a}_2, \ldots, \mathbf{a}_n^T \mathbf{a}_n$.

27. (a) **Solution.** $\langle 1, x \rangle = \int_{-1}^{1} 1x\, dx = \left. \frac{x^2}{2} \right|_{-1}^{1} = 0$

(c) **Hint.** Normalize 1 and x so as to make them unit vectors $u_1(x)$ and $u_2(x)$ and then determine the least squares solution in terms of $u_1(x)$ and $u_2(x)$.

29. Solution. In Example 3 it was shown that $\{ \frac{1}{\sqrt{2}}, \cos x, \ldots, \cos nx \}$ is an orthonormal set. In Section 4, Exercise 9 we saw that the functions $\cos kx$ and $\sin jx$ were orthogonal unit vectors in $C[-\pi, \pi]$. Furthermore

$$\left\langle \frac{1}{\sqrt{2}}, \sin jx \right\rangle = \frac{1}{\pi} \int_{-\pi}^{\pi} \frac{1}{\sqrt{2}} \sin jx\, dx = 0$$

and if $j \neq k$ then

$$\begin{aligned} \langle \sin jx, \sin kx \rangle &= \frac{1}{\pi} \int_{-\pi}^{\pi} \sin jx \sin kx\, dx \\ &= \frac{1}{2\pi} \int_{-\pi}^{\pi} (\cos(j-k)x - \cos(j+k)x)\, dx = 0 \end{aligned}$$

Therefore $\{1/\sqrt{2}, \cos x, \cos 2x, \ldots, \cos nx, \sin x, \sin 2x, \ldots, \sin nx\}$ is an orthonormal set of vectors.

33. Solution. Let

$$\mathbf{u}_i = \frac{1}{\|\mathbf{x}_i\|} \mathbf{x}_i \quad \text{for} \quad i = 1, \ldots, n$$

By Theorem 5.5.8 the best least squares approximation to $\mathbf{x}$ from S is given by

$$\mathbf{p} = \sum_{i=1}^{n} \langle \mathbf{x}, \mathbf{u}_i \rangle \mathbf{u}_i = \sum_{i=1}^{n} \frac{1}{\|\mathbf{x}_i\|^2} \langle \mathbf{x}, \mathbf{x}_i \rangle \mathbf{x}_i$$

$$= \sum_{i=1}^{n} \frac{\langle \mathbf{x}, \mathbf{x}_i \rangle}{\langle \mathbf{x}_i, \mathbf{x}_i \rangle} \mathbf{x}_i.$$

6 | THE GRAM-SCHMIDT ORTHOGONALIZATION PROCESS

Overview

In this section we learn a process for constructing an orthonormal basis for an n-dimensional inner product space V. The method involves using projections to transform an ordinary basis $\{\mathbf{x}_1, \mathbf{x}_2, \ldots, \mathbf{x}_n\}$ into an orthonormal basis $\{\mathbf{u}_1, \mathbf{u}_2, \ldots, \mathbf{u}_n\}$. When the method is applied to the column space of an $m \times n$ matrix we obtain a useful matrix factorization.

Important Concepts

1. **Gram-Schmidt Orthogonalization Process.** The *Gram-Schmidt process* is an algorithm for transforming an ordinary basis $\{\mathbf{x}_1, \mathbf{x}_2, \ldots, \mathbf{x}_n\}$ for an inner product space into an orthonormal basis $\{\mathbf{u}_1, \mathbf{u}_2, \ldots, \mathbf{u}_n\}$. The $\mathbf{u}_i$'s are constructed so that at each step of the process

   $$\text{Span}(\mathbf{u}_1, \mathbf{u}_2, \ldots, \mathbf{u}_k) = \text{Span}(\mathbf{x}_1, \mathbf{x}_2, \ldots, \mathbf{x}_k) \quad k = 1, \ldots, n$$

2. **Gram-Schmidt QR factorization.** If the Gram-Schmidt process is applied to the column space of an $m \times n$ matrix of rank n and you store the values of all the computed scalar products and norms in an upper triangular matrix R, then the matrix A can be factored into a product QR where Q is an $m \times n$ matrix whose columns form an orthonormal basis for $R(A)$.

3. **Modified Gram-Schmidt Process.** The *modified Gram-Schmidt process* is an alternative algorithm for performing the Gram-Schmidt process. The modified process has better numerical stability properties than the classical algorithm that you find in most linear algebra textbooks. If performed in exact arithmetic, the classical and modified methods will produce the same set of orthonormal vectors, however, when the computations are carried out in finite precision arithmetic, there can be a significant loss of orthogonality in the classical version due to roundoff error.

Important Theorems

▶ THEOREM 5.6.1 **(The Gram–Schmidt Process)** *Let* $\{\mathbf{x}_1, \mathbf{x}_2, \ldots, \mathbf{x}_n\}$ *be a basis for the inner product space* V. *Let*

$$\mathbf{u}_1 = \left(\frac{1}{\|\mathbf{x}_1\|}\right)\mathbf{x}_1$$

and define $\mathbf{u}_2, \ldots, \mathbf{u}_n$ *recursively by*

$$\mathbf{u}_{k+1} = \frac{1}{\|\mathbf{x}_{k+1} - \mathbf{p}_k\|}(\mathbf{x}_{k+1} - \mathbf{p}_k) \quad \text{for } k = 1, \ldots, n-1$$

where

$$\mathbf{p}_k = \langle \mathbf{x}_{k+1}, \mathbf{u}_1 \rangle \mathbf{u}_1 + \langle \mathbf{x}_{k+1}, \mathbf{u}_2 \rangle \mathbf{u}_2 + \cdots + \langle \mathbf{x}_{k+1}, \mathbf{u}_k \rangle \mathbf{u}_k$$

is the projection of $\mathbf{x}_{k+1}$ *onto* $\text{Span}(\mathbf{u}_1, \mathbf{u}_2, \ldots, \mathbf{u}_k)$. *The set*

$$\{\mathbf{u}_1, \mathbf{u}_2, \ldots, \mathbf{u}_n\}$$

is an orthonormal basis for V.

▶ THEOREM 5.6.2 **(QR Factorization)** *If* A *is an* $m \times n$ *matrix of rank* n, *then* A *can be factored into a product* QR, *where* Q *is an* $m \times n$ *matrix with orthonormal columns and* R *is an* $n \times n$ *matrix that is upper triangular and invertible.*

▶ THEOREM 5.6.3. *If* A *is an* $m \times n$ *matrix of rank* n, *then the solution to the least squares problem* $A\mathbf{x} = \mathbf{b}$ *is given by* $\hat{\mathbf{x}} = R^{-1}Q^T\mathbf{b}$, *where* Q *and* R *are the matrices obtained from the factorization given in Theorem 5.6.2. The solution* $\hat{\mathbf{x}}$ *may be obtained by using back substitution to solve* $R\mathbf{x} = Q^T\mathbf{b}$.

Exercises: Solutions and Hints

3. **Solution.** We will refer to the given basis vectors as $\mathbf{x}_1$, $\mathbf{x}_2$, and $\mathbf{x}_3$. At the first step of the process we compute the norm of $\mathbf{x}_1$

$$\|\mathbf{x}_1\| = \sqrt{1^2 + 2^2 + (-2)^2} = 3$$

and then normalize $\mathbf{x}_1$ to make it a unit vector.

$$\mathbf{u}_1 = \frac{1}{\|\mathbf{x}_1\|}\mathbf{x}_1 = \frac{1}{3}\mathbf{x}_1 = \begin{pmatrix} \frac{1}{3} \\ \frac{2}{3} \\ -\frac{2}{3} \end{pmatrix}$$

Next we compute the vector projection $\mathbf{p}_1$ of $\mathbf{x}_2$ onto $\mathbf{u}_1$

$$\mathbf{p}_1 = (\mathbf{u}_1^T\mathbf{x}_2)\mathbf{u}_1 = 2\mathbf{u}_1 = \begin{pmatrix} \frac{2}{3} \\ \frac{4}{3} \\ -\frac{4}{3} \end{pmatrix}$$

and subtract $\mathbf{p}_1$ from $\mathbf{x}_2$.

$$\mathbf{x}_2 - \mathbf{p}_1 = \begin{pmatrix} 4 \\ 3 \\ 2 \end{pmatrix} - \begin{pmatrix} \frac{2}{3} \\ \frac{4}{3} \\ -\frac{4}{3} \end{pmatrix} = \begin{pmatrix} \frac{10}{3} \\ \frac{5}{3} \\ \frac{10}{3} \end{pmatrix}$$

By construction the vector $\mathbf{x}_2 - \mathbf{p}_1$ is orthogonal to $\mathbf{u}_1$, so we need only normalize it to obtain our second orthonormal vector.

$$\|\mathbf{x}_2 - \mathbf{p}_1\| = \sqrt{\frac{100}{9} + \frac{25}{9} + \frac{100}{9}} = 5$$

$$\mathbf{u}_2 = \frac{1}{5}(\mathbf{x}_2 - \mathbf{p}_1) = \begin{pmatrix} \frac{2}{3} \\ \frac{1}{3} \\ \frac{2}{3} \end{pmatrix}$$

To compute the third orthonormal vector we project $\mathbf{x}_3$ onto $\mathrm{Span}(\mathbf{u}_1, \mathbf{u}_2)$

$$\mathbf{p}_2 = (\mathbf{u}_1^T \mathbf{x}_3)\mathbf{u}_1 + (\mathbf{u}_2^T \mathbf{x}_3)\mathbf{u}_2 = 1\mathbf{u}_1 + 2\mathbf{u}_2 = \begin{pmatrix} \frac{1}{3} \\ \frac{2}{3} \\ -\frac{2}{3} \end{pmatrix} + 2\begin{pmatrix} \frac{2}{3} \\ \frac{1}{3} \\ \frac{2}{3} \end{pmatrix} = \begin{pmatrix} \frac{5}{3} \\ \frac{4}{3} \\ \frac{2}{3} \end{pmatrix}$$

and then subtract the projection from $\mathbf{x}_3$.

$$\mathbf{x}_3 - \mathbf{p}_2 = \begin{pmatrix} 1 \\ 2 \\ 1 \end{pmatrix} - \begin{pmatrix} \frac{5}{3} \\ \frac{4}{3} \\ \frac{2}{3} \end{pmatrix} = \begin{pmatrix} -\frac{2}{3} \\ \frac{2}{3} \\ \frac{1}{3} \end{pmatrix}$$

Since $\mathbf{x}_3 - \mathbf{p}_2$ is a unit vector, it is not necessary to normalize. All we need do is set $\mathbf{u}_3 = \mathbf{x}_3 - \mathbf{p}_2$.

5. (a) **Solution.** At the first step of the Gram-Schmidt process we set

$$r_{11} = \|\mathbf{a}_1\| = \sqrt{2^2 + 1^2 + 2^2} = 3$$

and

$$\mathbf{q}_1 = \frac{1}{r_{11}}\mathbf{a}_1 = \begin{pmatrix} \frac{2}{3} \\ \frac{1}{3} \\ \frac{2}{3} \end{pmatrix}$$

Next we set $r_{12} = \mathbf{u}_1^T \mathbf{a}_2 = \frac{5}{3}$ and set

$$\mathbf{p}_1 = r_{12}\mathbf{q}_1 = \begin{pmatrix} \frac{10}{9} \\ \frac{5}{9} \\ \frac{10}{9} \end{pmatrix}$$

The vector $\mathbf{x}_2 - \mathbf{p}_1$ will be orthogonal to $\mathbf{q}_1$.

$$\mathbf{x}_2 - \mathbf{p}_1 = \begin{pmatrix} 1 \\ 1 \\ 1 \end{pmatrix} - \begin{pmatrix} \frac{10}{9} \\ \frac{5}{9} \\ \frac{10}{9} \end{pmatrix} = \begin{pmatrix} -\frac{1}{9} \\ \frac{4}{9} \\ -\frac{1}{9} \end{pmatrix}$$

Finally we set

$$r_{22} = \|\mathbf{x}_2 - \mathbf{p}_1\| = \sqrt{\left(-\frac{1}{9}\right)^2 + \left(\frac{4}{9}\right)^2 + \left(-\frac{1}{9}\right)^2} = \frac{\sqrt{2}}{3}$$

and

$$\mathbf{q}_2 = \frac{1}{r_{22}}(\mathbf{x}_2 - \mathbf{p}_1) = \begin{pmatrix} -\frac{\sqrt{2}}{6} \\ \frac{2\sqrt{2}}{3} \\ -\frac{\sqrt{2}}{6} \end{pmatrix}$$

10. **Solution.** Given a basis $\{x_1, \ldots, x_n\}$, one can construct an orthonormal basis using either the classical Gram–Schmidt process or the modified process. When carried out in exact arithmetic both methods will produce the same orthonormal set $\{\mathbf{q}_1, \ldots, \mathbf{q}_n\}$.

Proof: The proof is by induction on n. In the case $n = 1$, the vector $\mathbf{q}_1$ is computed in the same way for both methods.

$$\mathbf{q}_1 = \frac{1}{r_{11}}\mathbf{x}_1 \quad \text{where} \quad r_{11} = \|\mathbf{x}\|_1$$

Assume $\mathbf{q}_1, \ldots, \mathbf{q}_k$ are the same for both methods. In the classical Gram–Schmidt process one computes $\mathbf{q}_{k+1}$ as follows: Set

$$r_{i,k+1} = \langle \mathbf{x}_{k+1}, \mathbf{q}_i \rangle, \qquad i = 1, \ldots, k$$
$$\mathbf{p}_k = r_{1,k+1}\mathbf{q}_1 + r_{2,k+1}\mathbf{q}_2 + \cdots + r_{k,k+1}\mathbf{q}_k$$
$$r_{k+1,k+1} = \|\mathbf{x}_{k+1} - \mathbf{p}_k\|$$
$$\mathbf{q}_{k+1} = \frac{1}{r_{k+1,k+1}}(\mathbf{x}_{k+1} - \mathbf{p}_k)$$

Thus

$$\mathbf{q}_{k+1} = \frac{1}{r_{k+1,k+1}}(\mathbf{x}_{k+1} - r_{1,k+1}\mathbf{q}_1 - r_{2,k+1}\mathbf{q}_2 - \cdots - r_{k,k+1}\mathbf{q}_k)$$

In the modified version, at step 1 the vector $r_{1,k+1}\mathbf{q}_1$ is subtracted from $\mathbf{x}_{k+1}$.

$$\mathbf{x}_{k+1}^{(1)} = \mathbf{x}_{k+1} - r_{1,k+1}\mathbf{q}_1$$

At the next step $r_{2,k+1}\mathbf{q}_2$ is subtracted from $\mathbf{x}_{k+1}^{(1)}$.

$$\begin{aligned} \mathbf{x}_{k+1}^{(2)} &= \mathbf{x}_{k+1}^{(1)} - r_{2,k+1}\mathbf{q}_2 \\ &= \mathbf{x}_{k+1} - r_{1,k+1}\mathbf{q}_1 - r_{2,k+1}\mathbf{q}_2 \end{aligned}$$

In general after k steps we have

$$\begin{aligned} \mathbf{x}_{k+1}^{(k)} &= \mathbf{x}_{k+1} - r_{1,k+1}\mathbf{q}_1 - r_{2,k+1}\mathbf{q}_2 - \cdots - r_{k,k+1}\mathbf{q}_k \\ &= \mathbf{x}_{k+1} - \mathbf{p}_k \end{aligned}$$

In the last step we set

$$r_{k+1,k+1} = \|\mathbf{x}_{k+1}^{(k)}\| = \|\mathbf{x}_{k+1} - \mathbf{p}_k\|$$

and set

$$\mathbf{q}_{k+1} = \frac{1}{r_{k+1,k+1}}\mathbf{x}_{k+1}^{(k)} = \frac{1}{r_{k+1,k+1}}(\mathbf{x}_{k+1} - \mathbf{p}_k)$$

Thus $\mathbf{q}_{k+1}$ is the same as in the classical Gram–Schmidt process.

12. (a) **Hint.** Make use of Theorems 5.5.7 and 5.5.8.

   (b) **Hint.** Use the result from part (a).

   (c) **Hint.** The projection matrix corresponding to a subspace is unique.

14. **Hint.** If $\dim(U \cap V) = k > 0$, start with a basis $\{\mathbf{x}_1, \mathbf{x}_2, \ldots, \mathbf{x}_k\}$ for $U \cap V$ and extend it to a basis for U and also extend it to a basis for V.

7   ORTHOGONAL POLYNOMIALS

Overview

In this section we study families of orthogonal polynomials associated with various inner products on $C[a, b]$. Polynomials in each of these classes satisfy a three-term recursion relation. This recursion relation is particularly useful in computer applications. We also look at some classical families of orthogonal polynomials that have important applications in many areas.

Important Concepts

1. **Sequence of orthogonal polynomials.** Let $p_0(x), p_1(x), \ldots$ be a sequence of polynomials with $\deg p_i(x) = i$ for each i. If $\langle p_i(x), p_j(x) \rangle = 0$ whenever $i \neq j$, then $\{p_n(x)\}$ is said to be a *sequence of orthogonal polynomials*. If $\langle p_i, p_j \rangle = \delta_{ij}$, then $\{p_n(x)\}$ is said to be a *sequence of orthonormal polynomials*.

2. Legendre polynomials. The *Legendre polynomials* are orthogonal with respect to the inner product

$$\langle p, q \rangle = \int_{-1}^{1} p(x)q(x)\,dx$$

Let $P_n(x)$ denote the Legendre polynomial of degree n. If we choose the lead coefficients so that $P_n(1) = 1$ for each n, then the recursion formula for the Legendre polynomials is

$$(n+1)P_{n+1}(x) = (2n+1)xP_n(x) - nP_{n-1}(x)$$

3. Chebyshev polynomials. The *Chebyshev polynomials* are orthogonal with respect to the inner product,

$$\langle p, q \rangle = \int_{-1}^{1} p(x)q(x)(1-x^2)^{-1/2}\,dx$$

It is customary to normalize the lead coefficients so that $a_0 = 1$ and $a_k = 2^{k-1}$ for $k = 1, 2, \ldots$. The Chebyshev polynomials are denoted by $T_n(x)$ and have the interesting property that

$$T_n(\cos\theta) = \cos n\theta$$

Important Theorems

▶ **THEOREM 5.7.1.** *If $p_0, p_1, \ldots$ is a sequence of orthogonal polynomials, then*

(i) *$p_0, \ldots, p_{n-1}$ form a basis for P_n.*

(ii) *$p_n \in P_n^{\perp}$ (that is, p_n is orthogonal to every polynomial of degree less than n).*

▶ **THEOREM 5.7.2.** *Let $p_0, p_1, \ldots$ be a sequence of orthogonal polynomials. Let a_i denote the lead coefficient of p_i for each i and define $p_{-1}(x)$ to be the zero polynomial. Then*

$$\alpha_{n+1}p_{n+1}(x) = (x - \beta_{n+1})p_n(x) - \alpha_n\gamma_n p_{n-1}(x) \qquad (n \geq 0)$$

where $\alpha_0 = \gamma_0 = 1$ and

$$\alpha_n = \frac{a_{n-1}}{a_n}, \qquad \beta_n = \frac{\langle p_{n-1}, xp_{n-1}\rangle}{\langle p_{n-1}, p_{n-1}\rangle}, \qquad \gamma_n = \frac{\langle p_n, p_n\rangle}{\langle p_{n-1}, p_{n-1}\rangle} \qquad (n \geq 1)$$

▶ **THEOREM 5.7.3.** *If $p_0, p_1, p_2, \ldots$ is a sequence of orthogonal polynomials with respect to the inner product of the form*

$$\langle p, q \rangle = \int_{a}^{b} p(x)q(x)dx$$

then the zeros of $p_n(x)$ are all real and distinct and lie in the interval (a, b).

Exercises: Solutions and Hints

3. **Hint.** Let $x = \cos\theta$ and make use of the identity $T_n(x) = \cos n\theta$. This hint also applies to exercises 6, 7, and 8.

5. **Hint.** Write $p_n(x)$ in the form $p_n(x) = a_n x^n + q(x)$ where degree $q(x) < n$. By Theorem 5.7.1, $\langle q, p_n \rangle = 0$. It follows then that

$$\|p_n\|^2 = \langle p_n, p_n \rangle = \langle a_n x^n + q(x), p(x) \rangle$$

12. **Solution.** If $f(x)$ is a polynomial of degree less than n and $P(x)$ is the Lagrange interpolating polynomial that agrees with $f(x)$ at $x_1, \ldots, x_n$, then degree $P(x) \leq n - 1$. If we set

$$h(x) = P(x) - f(x)$$

then the degree of h is also $\leq n - 1$ and

$$h(x_i) = P(x_i) - f(x_i) = 0 \qquad i = 1, \ldots, n$$

Therefore h must be the zero polynomial and hence

$$P(x) = f(x)$$

15. (a) **Hint.** If $f(x)$ is a polynomial of degree less than n, then the Lagrange polynomial that interpolates $f(x)$ at $x_1, x_2, \ldots, x_n$ must be equal to $f(x)$.

17. (a) **Hint.** What will the γ_j's turn out to be in this case?

MATLAB EXERCISES

3. (c) From the graph it should be clear that you get a better fit at the bottom of the atmosphere.

5. (a) A is the product of two random matrices. One would expect that both of the random matrices will have full rank, that is, rank 2. Since the row vectors of A are linear combinations of the row vectors of the second random matrix, one would also expect that A would have rank 2. If the rank of A is 2, then the nullity of A should be $5 - 2 = 3$.

(b) Since the column vectors of Q form an orthonormal basis for $R(A)$ and the column vectors of W form an orthonormal basis for $N(A^T) = R(A)^\perp$, the column vectors of $S = (Q \ \ W)$ form an orthonormal basis for R^5 and hence S is an orthogonal matrix. Each column vector of W is in $N(A^T)$. thus it follows that

$$A^T W = O$$

and

$$W^T A = (A^T W)^T = O^T$$

(c) See the hint given in the book.

(d) **Hint.** If $\mathbf{b} \in R(A)$, then $\mathbf{b} = A\mathbf{x}$ for some $\mathbf{x} \in R^5$. Make use of the result from part (c) to show $QQ^T\mathbf{b} = \mathbf{b}$.

(f) Since the projection of a vector onto a subspace is unique, $\mathbf{w}$ must equal $\mathbf{r}$.

CHAPTER TEST A

1. **Hint.** What if your two vectors are orthogonal?
2. **Hint.** If $|\mathbf{x}^T\mathbf{y}| = 1$, what can you conclude about the angle between the vectors?
3. **Hint.** Look at Exercise 13 of Section 1. If you can find vectors $\mathbf{x}_1$, $\mathbf{x}_2$, $\mathbf{x}_3$ such that $\mathbf{x}_1 \perp \mathbf{x}_2$ and $\mathbf{x}_2 \perp \mathbf{x}_3$, but $\mathbf{x}_1$ is not orthogonal to $\mathbf{x}_3$, then consider the subspaces

$$S_1 = \text{Span}(\mathbf{x}_1), \quad S_2 = \text{Span}(\mathbf{x}_2), \quad S_3 = \text{Span}(\mathbf{x}_3)$$

4. **Hint.** If $A^T\mathbf{y} = \mathbf{0}$, then $\mathbf{y}$ is in $N(A^T)$.
5. **Hint.** The matrices A and A^TA have the same rank. (See Exercise 13 of Section 2.)
6. **Hint.** The least squares problem will not have a unique solution but that doesn't imply that the projection is not unique. See Theorem 5.3.1.
7. **Hint.** If A is $m \times n$ and $N(A) = \{\mathbf{0}\}$, then what can you conclude about the rank of A?
8. **Hint.** Check to see if $(Q_1Q_2)^T(Q_1Q_2) = I$.
9. **Hint.** How is the (i, j) entry of U^TU determined?
10. **Hint.** Make up some examples (with $k < n$) to see if the statement is true.

CHAPTER TEST B

4. (b) **Hint.** The vector $\mathbf{x}$ is in both $R(A)$ and $N(A^T)$.
5. **Hint.** If ϕ is the angle between $Q\mathbf{x}$ and $Q\mathbf{y}$ and θ is the angle between $\mathbf{x}$ and $\mathbf{y}$, show that $\cos\phi = \cos\theta$.
9. (a) **Hint.** Show that if $\mathbf{q}_j$ is the jth column vector of Q then $\mathbf{q}_j$ is in $N(A^T)$ and show that $P\mathbf{x} = \mathbf{0}$ for any vector $\mathbf{x}$ in $N(A^T)$.
   (b) **Hint.** If $\{\mathbf{u}_1, \mathbf{u}_2, \mathbf{u}_3, \mathbf{u}_4\}$ is an orthonormal basis for $R(A)$ and $\{\mathbf{u}_5, \mathbf{u}_6, \mathbf{u}_7\}$ is an orthonormal basis for $N(A^T)$ and we set

$$U_1 = (\mathbf{u}_1, \mathbf{u}_2, \mathbf{u}_3, \mathbf{u}_4) \qquad U_2 = (\mathbf{u}_5, \mathbf{u}_6, \mathbf{u}_7)$$

then $U = (U_1, U_2)$ is an orthogonal matrix.
11. (b) **Hint.** The vectors $\cos x$ and $\sin x$ are orthogonal, so you can use the Pythagorean Law.

Chapter 6

genvalues

NVECTORS

ral concepts of linear algebra. Eigenvalues
ociated with linear transformations. If A
epresenting a linear transformation from
rs provide the key to understanding how
vide a natural coordinate system for the
s are the scaling factors for the system.
itions and properties of eigenvalues and
mples.

Let A be an $n \times n$ matrix. A scalar λ is
teristic value of A if there exists a nonzero
vector $\mathbf{x}$ is said to be an *eigenvector* or a
λ.

characteristic equation.
ain an nth-degree polynomial in the vari-

$\det(A - \lambda I)$

characteristic polynomial, and equation
ic equation for the matrix A. The roots
e the eigenvalues of A. If we count roots
acteristic polynomial will have exactly n

f a matrix A, then $N(A - \lambda I)$ is referred to
λ. Every nonzero vector in the eigenspace

he diagonal entries of A is called the *trace*

Important Theorems and Results

▶ **(Equivalent conditions)** Let A be an $n \times n$ matrix and λ be a scalar. The following statements are equivalent:

(a) λ is an eigenvalue of A.

(b) $(A - \lambda I)\mathbf{x} = \mathbf{0}$ has a nontrivial solution.

(c) $N(A - \lambda I) \neq \{\mathbf{0}\}$

(d) $A - \lambda I$ is singular.

(e) $\det(A - \lambda I) = 0$

▶ **(Complex Eigenvalues and Eigenvectors)** Let A be an $n \times n$ matrix whose entries are all real numbers. If $\lambda = a + bi$ $(b \neq 0)$ is an eigenvalue of A, then $\overline{\lambda} = a - bi$ must also be an eigenvalue of A. Furthermore, if $\mathbf{x}$ is an eigenvector belonging to the complex eigenvalue λ, then $\overline{\mathbf{x}}$ is an eigenvector belonging to $\overline{\lambda}$.

▶ **(The Product and Sum of the Eigenvalues)** If A is an $n \times n$ matrix with eigenvalues $\lambda_1, \ldots, \lambda_n$, then

$$\det(A) = \lambda_1 \cdot \lambda_2 \cdots \lambda_n$$

and

$$\operatorname{tr}(A) = \sum_{i=1}^{n} \lambda_i$$

▶ THEOREM 6.1.1 **(Eigenvalues of Similar Matrices)** *Let A and B be $n \times n$ matrices. If B is similar to A, then the two matrices both have the same characteristic polynomial and consequently both have the same eigenvalues.*

Exercises: Solutions and Hints

1. (i) **Solution.** To determine the characteristic polynomial we compute the cofactor expansion of $\det(A - \lambda I)$ along the first column

$$p(\lambda) = \begin{pmatrix} 4 - \lambda & -5 & 1 \\ 1 & -\lambda & -1 \\ 0 & 1 & -1 - \lambda \end{pmatrix}$$

$$= (4 - \lambda)(\lambda^2 + \lambda + 1) - (5 + 5\lambda - 1)$$
$$= -\lambda^3 + 3\lambda^2 + 3\lambda + 4 - 4 - 5\lambda$$
$$= -\lambda^3 + 3\lambda^2 - 2\lambda$$
$$= -\lambda(\lambda - 2)(\lambda - 1)$$

The eigenvalues are the roots of the characteristic equation

$$-\lambda(\lambda - 2)(\lambda - 1) = 0$$

Thus $\lambda_1 = 0$, $\lambda_2 = 2$ and $\lambda_3 = 1$ are the eigenvalues of the matrix A. To determine the eigenspace corresponding to $\lambda_1 = 0$ we must find a basis for $N(A - 0I) = N(A)$. To do this we compute the reduced row echelon form of $(A \,|\, \mathbf{0})$.

$$
\left(\begin{array}{ccc|c}
4 & -5 & 1 & 0 \\
1 & 0 & -1 & 0 \\
0 & 1 & -1 & 0
\end{array}\right)
\rightarrow
\left(\begin{array}{ccc|c}
1 & 0 & -1 & 0 \\
0 & 1 & -1 & 0 \\
0 & 0 & 0 & 0
\end{array}\right)
$$

The variable x_3 is free and $x_1 = x_3$, $x_2 = x_3$. It follows that the eigenvectors belonging to $\lambda_1 = 0$ are all of the form $(x_3, x_3, x_3)^T = x_3(1,1,1)^T$ So the eigenspace is spanned by the vector $\mathbf{x}_1 = (1,1,1)^T$.

To find a basis for the eigenspace corresponding to $\lambda_2 = 2$ we compute the reduced row echelon form of $(A - 2I \,|\, \mathbf{0})$.

$$
\left(\begin{array}{ccc|c}
2 & -5 & 1 & 0 \\
1 & -2 & -1 & 0 \\
0 & 1 & -3 & 0
\end{array}\right)
\rightarrow
\left(\begin{array}{ccc|c}
1 & 0 & -7 & 0 \\
0 & 1 & -3 & 0 \\
0 & 0 & 0 & 0
\end{array}\right)
$$

The variable x_3 is free and $x_1 = 7x_3$, $x_2 = 3x_3$. It follows that the eigenvectors belonging to $\lambda_2 = 2$ are all of the form $(7x_3, 3x_3, x_3)^T = x_3(7,3,1)^T$ So the eigenspace is spanned by the vector $\mathbf{x}_2 = (7,3,1)^T$.

To find a basis for the eigenspace corresponding to $\lambda_3 = 1$ we compute the reduced row echelon form of $(A - I \,|\, \mathbf{0})$.

$$
\left(\begin{array}{ccc|c}
3 & -5 & 1 & 0 \\
1 & -1 & -1 & 0 \\
0 & 1 & -2 & 0
\end{array}\right)
\rightarrow
\left(\begin{array}{ccc|c}
1 & 0 & -3 & 0 \\
0 & 1 & -2 & 0 \\
0 & 0 & 0 & 0
\end{array}\right)
$$

The variable x_3 is free and $x_1 = 3x_3$, $x_2 = 2x_3$. It follows that the eigenvectors belonging to $\lambda_3 = 1$ are all of the form $(3x_3, 2x_3, x_3)^T = x_3(3,2,1)^T$ So the eigenspace is spanned by the vector $\mathbf{x}_3 = (3,2,1)^T$.

3. **Hint.** $\det(A) = \det(A - 0I)$.

4. **Hint.** Make use of the eigenvector $\mathbf{x}$ belonging to λ.

6. **Hint.** If λ is an eigenvalue of A and $\mathbf{x}$ is an eigenvector belonging to λ, then show that $(\lambda^2 - \lambda)\mathbf{x} = \mathbf{0}$.

7. **Hint.** If λ is an eigenvalue of A, then by Exercise 5, λ^k is also an eigenvalue of A^k.

9. **Hint.** Look at some nonsymmetric 2×2 examples.

10. **Hint.** See Example 2 in Section 2 of Chapter 4.

15. **Hint.** Use the Rank-Nullity Theorem.

17. (a) **Solution.** If $\alpha = a + bi$ and $\beta = c + di$, then
$$\overline{\alpha + \beta} = \overline{(a+c) + (b+d)i} = (a+c) - (b+d)i$$
and
$$\overline{\alpha} + \overline{\beta} = (a - bi) + (c - di) = (a+c) - (b+d)i$$
Therefore $\overline{\alpha + \beta} = \overline{\alpha} + \overline{\beta}$.

Next we show that the conjugate of the product of two numbers is the product of the conjugates.
$$\overline{\alpha\beta} = \overline{(ac - bd) + (ad + bc)i} = (ac - bd) - (ad + bc)i$$
$$\overline{\alpha}\,\overline{\beta} = (a - bi)(c - di) = (ac - bd) - (ad + bc)i$$
Therefore $\overline{\alpha\beta} = \overline{\alpha}\,\overline{\beta}$.

(b) **Hint.** Show that the (i, j) entries of $\overline{AB}$ and $\overline{A}\,\overline{B}$ are equal.

18. **Hint.** Let $\mathbf{x}$ be an eigenvector of Q. Since Q is an orthogonal matrix we have $\|\mathbf{x}\| = \|Q\mathbf{x}\|$.

20. (b) **Hint.** First show that λ_2 must be a scalar of the form $\cos\theta + i\sin\theta$ for some angle θ.

22. **Hint.** You need to show that $\mathbf{y} = S\mathbf{x}$ is nonzero and that $A\mathbf{y} = \lambda\mathbf{y}$.

23. **Hint.** If $A\mathbf{x} = \lambda_1\mathbf{x}$ and $B\mathbf{x} = \lambda_2\mathbf{x}$, then show that $C\mathbf{x}$ will equal a scalar multiple of $\mathbf{x}$.

24. **Hint.** If $\lambda \neq 0$ and $\mathbf{x}$ is an eigenvector belonging to λ, then $\mathbf{x} = \frac{1}{\lambda}A\mathbf{x}$.

26. **Hint.** If the columns of A each add up to a fixed constant δ then the row vectors of $A - \delta I$ all add up to $(0, 0, \ldots, 0)$.

27. **Hint.** Consider the expression $\mathbf{x}^T A^T \mathbf{y}$ and note that it can also be written in the form $(A\mathbf{x})^T\mathbf{y}$.

28. (a) **Hint.** Let $\mathbf{x}$ be and eigenvector of AB belonging to λ (where $\lambda \neq 0$) and let $\mathbf{y} = B\mathbf{x}$. Show that $\mathbf{y} \neq \mathbf{0}$.

(b) **Hint.** If $\lambda = 0$ is an eigenvalue of AB, then AB must be singular.

---

2 | SYSTEMS OF LINEAR DIFFERENTIAL EQUATIONS

Overview

Eigenvalues play an important role in the solution of systems of linear differential equations. In this section we see how they are used in the solution of systems of linear differential equations with constant coefficients.

Important Concepts

1. **Vector functions and their derivatives.** If $y_1(t), \ldots, y_n(t)$ are differentiable functions of t, then the vector function $\mathbf{Y}(t)$ and its derivative $\mathbf{Y}'(t)$

are defined by

$$
\mathbf{Y}(t) = \begin{pmatrix} y_1(t) \\ y_2(t) \\ \vdots \\ y_n(t) \end{pmatrix}, \quad
\mathbf{Y}'(t) = \begin{pmatrix} y_1'(t) \\ y_2'(t) \\ \vdots \\ y_n'(t) \end{pmatrix}
$$

2. **Linear system of differential equations.** A *constant coefficient linear first order system of differential equations* is a system of the form $\mathbf{Y}' = A\mathbf{Y}$. A *solution* is a vector function $\mathbf{Y}(t)$ that works in the equation.

3. **Initial value problem.** A problem of the form

$$
\mathbf{Y}' = A\mathbf{Y}, \qquad \mathbf{Y}(0) = \mathbf{Y}_0
$$

is called an *initial value problem*.

Important Results

▶ If λ is an eigenvalue of A and $\mathbf{x}$ is an eigenvector belonging to λ, then $\mathbf{Y}(t) = e^{\lambda t}\mathbf{x}$ is a solution to the system $\mathbf{Y}' = A\mathbf{Y}$.

▶ If A is an $n \times n$ matrix with eigenvalues $\lambda_1, \lambda_2, \ldots, \lambda_n$ and linearly independent eigenvectors $\mathbf{x}_1, \mathbf{x}_2, \ldots, \mathbf{x}_n$, then the general solution to the system $\mathbf{Y}' = A\mathbf{Y}$ is given by

$$
\mathbf{Y}(t) = c_1 e^{\lambda_1 t}\mathbf{x}_1 + c_1 e^{\lambda_1 t}\mathbf{x}_2 + \cdots c_n e^{\lambda_n t}\mathbf{x}_n
$$

To find the coefficients $c_1, c_2, \ldots, c_n$ of the solution to the initial value problem

$$
\mathbf{Y}' = A\mathbf{Y}, \qquad \mathbf{Y}(0) = \mathbf{Y}_0
$$

one must solve the linear system $X\mathbf{c} = \mathbf{Y}_0$ where $X = (\mathbf{x}_1, \mathbf{x}_2, \ldots, \mathbf{x}_n)$.

Exercises: Solutions and Hints

2. (a) **Solution.** The eigenvalues of the coefficient matrix are $\lambda_1 = -3$ and $\lambda_2 = 1$. The corresponding eigenvectors are

$$
\mathbf{x}_1 = \begin{pmatrix} -1 \\ 1 \end{pmatrix} \quad \text{and} \quad \mathbf{x}_2 = \begin{pmatrix} 1 \\ 1 \end{pmatrix}
$$

The general solution to the system without initial conditions is

$$
\mathbf{Y} = c_1 e^{-3t} \begin{pmatrix} -1 \\ 1 \end{pmatrix} + c_2 e^t \begin{pmatrix} 1 \\ 1 \end{pmatrix} = \begin{pmatrix} -c_1 e^{-3t} + c_2 e^t \\ c_1 e^{-3t} + c_2 e^t \end{pmatrix}
$$

To find the solution to the initial value problem we must solve the system $X\mathbf{c} = \mathbf{Y}_0$. But this is simply the system

$$-c_1 + c_2 = 3$$
$$c_1 + c_2 = 1$$

The solution $c_1 = -1$, $c_2 = 2$ is easy to find. Using these values it follows that the solution to the initial value problem is

$$\mathbf{Y} = -e^{-3t}\begin{pmatrix} -1 \\ 1 \end{pmatrix} + 2e^t \begin{pmatrix} 1 \\ 1 \end{pmatrix} = \begin{pmatrix} e^{-3t} + 2e^t \\ -e^{-3t} + 2e^t \end{pmatrix}$$

9. **Solution.** If x_1 and x_2 represent the displacements of the two masses from the equilibrium positions, then by Hooke's law we have forces of $k_1 x_1$ and $-k_2(x_2 - x_1)$ acting on the first mass. We also have a force $-m_1 g$ due to gravity acting on the mass. By Newton's second law of motion we have

$$m_1 x_1'' = k_1 x_1 - k_2(x_2 - x_1) - m_1 g$$

Similarly for the second mass we have a force $k_2(x_2 - x_1)$ (by Hooke's law) and a force $-m_2 g$ (gravity). So by Newton's law

$$m_2 x_2'' = k_2(x_2 - x_1) - m_2 g$$

Therefore the motion can be modeled by the second order system

$$m_1 x_1'' = k_1 x_1 - k_2(x_2 - x_1) - m_1 g$$
$$m_2 x_2'' = k_2(x_2 - x_1) - m_2 g$$

11. **Partial Solution and Hint.** If

$$y^{(n)} = a_0 y + a_1 y' + \cdots + a_{n-1} y^{(n-1)}$$

and we set

$$y_1 = y, \ y_2 = y_1' = y'', \ y_3 = y_2' = y''', \ldots, y_n = y_{n-1}' = y^n$$

then the nth order equation can be written as a system of first order equations of the form $\mathbf{Y}' = A\mathbf{Y}$ where

$$A = \begin{pmatrix} 0 & 1 & 0 & \cdots & 0 \\ 0 & 0 & 1 & \cdots & 0 \\ \vdots & & & & \\ 0 & 0 & 0 & \cdots & 1 \\ a_0 & a_1 & a_2 & \cdots & a_{n-1} \end{pmatrix}$$

The easiest way to determine the characteristic polynomial is to do a cofactor expansion of $\det(A - \lambda I)$ along the last row of the matrix.

3 | DIAGONALIZATION

Overview

If an $n \times n$ matrix A has n linearly independent eigenvectors, then the eigenvectors are an ideal basis to use for applications involving linear transformations represented by A. In fact the matrix representation of the linear transformation $L(\mathbf{x}) = A\mathbf{x}$ with respect to a basis of eigenvectors will be a diagonal matrix. The change of basis and the diagonal representation can both be expressed in terms of a factorization of A. In this section we learn a necessary and sufficient condition for the existence of such a factorization and look at a number of examples. If the factorization is not possible, an alternative that works well in most applications is to use a special function of A known as the matrix exponential.

Important Concepts

1. **Diagonalizable matrices.** An $n \times n$ matrix A is said to be *diagonalizable* if there exists a nonsingular matrix X and a diagonal matrix D such that

$$X^{-1}AX = D$$

We say that X *diagonalizes* A.

2. **Defective matrices.** If an $n \times n$ matrix A has fewer than n linearly independent eigenvectors, we say that A is *defective*.

3. **Stochastic process.** A *stochastic process* is any sequence of experiments for which the outcome at any stage depends on chance.

4. **Markov Process.** A *Markov process* is a stochastic process with the following properties:
   (i) The set of possible outcomes or states is finite.
   (ii) The probability of the next outcome depends only on the previous outcome.
   (iii) The probabilities are constant over time.

5. **Probability vector** A vector $\mathbf{x}$ in R^n is said to be a *probability vector* if it entries are nonnegative and add up to 1.

6. **Markov chain.** Let A be a matrix representing a Markov process in the sense that the (i, j) entry of A is equal to the probability that if the previous outcome of the process was state i then the next outcome will be state j. Given an initial probability vector $\mathbf{x}_0$, we can generate a sequence of vectors by setting

$$\mathbf{x}_{n+1} = A\mathbf{x}_n \quad \text{for} \quad n = 0, 1, 2, \ldots$$

The vectors $\mathbf{x}_i$ produced in this manner are referred to as *state vectors* and the sequence of state vectors is called a *Markov chain*. The matrix A is referred to as a transition matrix for the Markov process.

7. **Matrix exponential.** Given an $n \times n$ matrix A, the *matrix exponential* e^A is defined by the convergent power series

$$e^A = I + A + \frac{1}{2!}A^2 + \frac{1}{3!}A^3 + \cdots$$

Important Theorems

▶ **THEOREM 6.3.1.** *If* $\lambda_1, \lambda_2, \ldots, \lambda_k$ *are distinct eigenvalues of an* $n \times n$ *matrix* A *with corresponding eigenvectors* $\mathbf{x}_1, \mathbf{x}_2, \ldots, \mathbf{x}_k$, *then* $\mathbf{x}_1, \ldots, \mathbf{x}_k$ *are linearly independent.*

▶ **THEOREM 6.3.2.** *An* $n \times n$ *matrix* A *is diagonalizable if and only if* A *has* n *linearly independent eigenvectors.*

▶ **THEOREM 6.3.3.** *If a Markov chain with an* $n \times n$ *transition matrix* A *converges to a steady-state vector* $\mathbf{x}$, *then*

(i) $\mathbf{x}$ *is a probability vector.*

(ii) $\lambda_1 = 1$ *is an eigenvalue of* A *and* $\mathbf{x}$ *is an eigenvector belonging to* λ_1.

▶ **THEOREM 6.3.4.** *If* $\lambda_1 = 1$ *is a dominant eigenvalue of a stochastic matrix* A, *then the Markov chain with transition* A *will converge to a steady-state vector.*

Exercises: Solutions and Hints

1. **Solution.** The factorization XDX^{-1} is not unique. However the diagonal elements of D must be eigenvalues of A and if λ_i is the ith diagonal element of D, then $\mathbf{x}_i$ must be an eigenvector belonging to λ_i.

   (e) The eigenvalues of A are $\lambda_1 = 1$, $\lambda_2 = 2$, $\lambda_3 = -2$ and the corresponding eigenvectors are

   $$
   \mathbf{x}_1 = \begin{pmatrix} 3 \\ 1 \\ 2 \end{pmatrix}, \quad
   \mathbf{x}_2 = \begin{pmatrix} 0 \\ 3 \\ 1 \end{pmatrix}, \quad
   \mathbf{x}_3 = \begin{pmatrix} 0 \\ -1 \\ 1 \end{pmatrix}
   $$

   If we set $X = (\mathbf{x}_1, \mathbf{x}_2, \mathbf{x}_3)$ and we compute its inverse we can then factor A as follows

   $$
   A = XDX^{-1} =
   \begin{pmatrix} 3 & 0 & 0 \\ 1 & 3 & -1 \\ 2 & 1 & 1 \end{pmatrix}
   \begin{pmatrix} 1 & 0 & 0 \\ 0 & 2 & 0 \\ 0 & 0 & -2 \end{pmatrix}
   \begin{pmatrix} \frac{1}{3} & 0 & 0 \\ -\frac{1}{4} & \frac{1}{4} & \frac{1}{4} \\ -\frac{5}{12} & -\frac{1}{4} & \frac{3}{4} \end{pmatrix}
   $$

2. (e) **Solution.** If $A = XDX^{-1}$, then $A^6 = XD^6X^{-1}$. Using the XDX^{-1} factorization from Exercise 1(e), we have

   $$
   A^6 =
   \begin{pmatrix} 3 & 0 & 0 \\ 1 & 3 & -1 \\ 2 & 1 & 1 \end{pmatrix}
   \begin{pmatrix} 1 & 0 & 0 \\ 0 & 2 & 0 \\ 0 & 0 & -2 \end{pmatrix}^{6}
   \begin{pmatrix} \frac{1}{3} & 0 & 0 \\ -\frac{1}{4} & \frac{1}{4} & \frac{1}{4} \\ -\frac{5}{12} & -\frac{1}{4} & \frac{3}{4} \end{pmatrix}
   $$

$$= \begin{pmatrix} 1 & 0 & 0 \\ -21 & 64 & 0 \\ -42 & 0 & 64 \end{pmatrix}$$

3. (e) **Solution.** If $A = XDX^{-1}$ is nonsingular, then $A^{-1} = XD^{-1}X^{-1}$. Using the XDX^{-1} factorization from Exercise 1(e), we have

$$A^{-1} = \begin{pmatrix} 3 & 0 & 0 \\ 1 & 3 & -1 \\ 2 & 1 & 1 \end{pmatrix} \begin{pmatrix} 1 & 0 & 0 \\ 0 & 2 & 0 \\ 0 & 0 & -2 \end{pmatrix}^{-1} \begin{pmatrix} \frac{1}{3} & 0 & 0 \\ -\frac{1}{4} & \frac{1}{4} & \frac{1}{4} \\ -\frac{5}{12} & -\frac{1}{4} & \frac{3}{4} \end{pmatrix}$$

$$= \begin{pmatrix} 1 & 0 & 0 \\ -\frac{1}{4} & \frac{1}{4} & \frac{3}{4} \\ \frac{3}{4} & \frac{1}{4} & -\frac{1}{4} \end{pmatrix}$$

4. Hint. If A has an XDX^{-1} factorization, can you find a diagonal matrix D_1 such that $D_1^2 = D$?

6. Hint. If $A = XDX^{-1}$ and the diagonal entries of D are all 1 or -1, show that $D^{-1} = D$.

7. Hint. Show that the eigenspace corresponding to the eigenvalue a has dimension 1.

8. Hint. If A has distinct eigenvalues then it is diagonalizable. If A has multiple eigenvalues then it may or may not be defective depending on the dimensions on the eigenspaces.

9. Hint. Make use of the Rank-Nullity theorem.

11. Hint. Make use of the Rank-Nullity theorem.

13. Hint. Show that any vector in the column space of A can be written as a linear combination of the eigenvectors of A that correspond to the nonzero eigenvalues.

15. (a) **Hint.** A and B are similar, so they have the same eigenvalues $\lambda_1, \lambda_2, \ldots, \lambda_n$. Here we are assuming

$$\lambda_1 = \lambda_2 = \cdots = \lambda_k = \lambda$$

Show that for the matrix B, the vector $\mathbf{e}_i$ is an eigenvector of λ_i for $i = 1, \ldots, k$.

16. (a) **Hint.** Make use of the Rank-Nullity theorem.

(b) **Hint.** How is $\operatorname{tr} A$ related to the eigenvalues of A?

17. **Hint.** If B is similar to A and $A = XDX^{-1}$, find a matrix Y such that $B = YDY^{-1}$.

18. **Hint.** If A and B both have the same diagonalizing matrix X, then

$$A = XD_1X^{-1} \qquad \text{and} \qquad B = XD_2X^{-1}$$

19. **Hint.** If $\mathbf{r}_j$ is the eigenvector of T corresponding to the eigenvalue $\lambda_j = t_{jj}$, show that its last $n - j$ entries are all equal to 0.

24. **Hint.** Use Theorems 6.3.3 and 6.3.4 and make use of the result from Exercise 23.

26. (c) **Hint.** Show that for $k = 1, 2, \ldots$

$$A^k = \begin{pmatrix} 1 & 0 & -k \\ 0 & 1 & 0 \\ 0 & 0 & 1 \end{pmatrix}$$

27. **Hint.** These matrices are diagonalizable so in each case

$$e^A = Xe^DX^{-1}$$

28. (d) **Hint.** The matrix A is defective, so e^{tA} must be computed using the definition of the matrix exponential.

29. **Hint.** If λ is an eigenvalue of A and $\mathbf{x}$ is an eigenvector belonging to λ then

$$e^A\mathbf{x} = \left(I + A + \frac{1}{2!}A^2 + \frac{1}{3!}A^3 + \cdots \right)\mathbf{x}$$

30. **Hint.** If A is diagonalizable then how are the eigenvalues of e^A related to the eigenvalues of A?

4 | HERMITIAN MATRICES

Overview

Let C^n denote the vector space of all n-tuples of complex numbers. The set C of all complex numbers will be taken as our field of scalars. We have already seen that a matrix A with real entries may have complex eigenvalues and eigenvectors. In this section we study matrices with complex entries and look at the complex analogues of symmetric and orthogonal matrices. The complex analogues of symmetric matrices are called Hermitian matrices. The eigenvalue and eigenvector theory is particularly nice for this class of matrices.

Important Concepts

1. **Lengths of complex scalars and vectors.** If $\alpha = a + bi$ is a complex scalar, the *length* of α is given by

$$|\alpha| = \sqrt{\bar{\alpha}\alpha} = \sqrt{a^2 + b^2}$$

The *length of a vector* $\mathbf{z} = (z_1, z_2, \ldots, z_n)^T$ in C^n is given by

$$\|\mathbf{z}\| = \left(\overline{\mathbf{z}}^T \mathbf{z}\right)^{1/2} = \left(\mathbf{z}^H \mathbf{z}\right)^{1/2}$$

where $\mathbf{z}^H$ is the vector formed from $\mathbf{z}$ by conjugating all of its entries and then transposing the conjugated vector.

2. **Complex inner product.**   Let V be a vector space over the complex numbers. An *inner product* on V is an operation that assigns to each pair of vectors $\mathbf{z}$ and $\mathbf{w}$ in V a complex number $\langle \mathbf{z}, \mathbf{w} \rangle$ satisfying the following conditions.
   (i) $\langle \mathbf{z}, \mathbf{z} \rangle \geq 0$ with equality if and only if $\mathbf{z} = \mathbf{0}$.
   (ii) $\langle \mathbf{z}, \mathbf{w} \rangle = \overline{\langle \mathbf{w}, \mathbf{z} \rangle}$ for all $\mathbf{z}$ and $\mathbf{w}$ in V.
   (iii) $\langle \alpha\mathbf{z} + \beta\mathbf{w}, \mathbf{u} \rangle = \alpha\langle \mathbf{z}, \mathbf{u} \rangle + \beta\langle \mathbf{w}, \mathbf{u} \rangle$.

3. **Inner product on C^n.**   We can define an inner product on C^n by

$$\langle \mathbf{z}, \mathbf{w} \rangle = \mathbf{w}^H \mathbf{z}$$

   for all $\mathbf{z}$ and $\mathbf{w}$ in C^n.

4. **Notation for the conjugate transpose of a matrix.**   If A is a matrix with complex entries, then A^H is the matrix formed by conjugating each the entries of A and then transposing the matrix. The (i, j) entry of A^H is $\overline{a_{ji}}$.

5. **Hermitian matrices.**   A matrix M is said to be *Hermitian* if $M = M^H$.

6. **Unitary matrices.**   An $n \times n$ matrix U is said to be *unitary* if its column vectors form an orthonormal set in C^n.

7. **Schur decomposition.**   Every $n \times n$ matrix A can be factored into a product UTU^H, where U is unitary and T is upper triangular. The UTU^H factorization is called the *Schur decomposition* of A.

8. **Normal matrices.**   A matrix A is said to be *normal* if $AA^H = A^HA$.

Important Theorems and Results

▶ **(Properties of the inner product on C^n)**

$$\langle \mathbf{z}, \mathbf{w} \rangle = \mathbf{w}^H \mathbf{z}$$
$$\mathbf{z}^H \mathbf{w} = \overline{\mathbf{w}^H \mathbf{z}}$$
$$\|\mathbf{z}\|^2 = \mathbf{z}^H \mathbf{z}$$

▶ THEOREM 6.4.1. *The eigenvalues of a Hermitian matrix are all real. Furthermore, eigenvectors belonging to distinct eigenvalues are orthogonal.*

▶ THEOREM 6.4.3 **(Schur's Theorem)** *For each $n \times n$ matrix A, there exists a unitary matrix U such that U^HAU is upper triangular.*

▶ THEOREM 6.4.4 **(Spectral Theorem)** *If A is Hermitian, then there exists a unitary matrix U that diagonalizes A.*

▶ COROLLARY 6.4.5. *If A is a real symmetric matrix, then there is an orthogonal matrix U that diagonalizes A, that is, $U^TAU = D$, where D is diagonal.*

▶ THEOREM 6.4.6. *A matrix A is normal if and only if A possesses a complete orthonormal set of eigenvectors.*

Exercises: Solutions and Hints

2. (a) **Hint.** Show that

$$z_2^H z_1 = 0 \quad \text{and} \quad z_1^H z_1 = z_2^H z_2 = 1$$

(b) **Hint.** Note for a real inner product space V with an orthonormal basis $\{u_1, \ldots, u_n\}$, we have by Theorem 5.5.2 that if

$$x = c_1 u_1 + \cdots + c_n u_n$$

then

$$c_j = \langle u_j, x \rangle = \langle x, u_j \rangle$$

For the complex case if $\{u_1, \ldots, u_n\}$ is an orthonormal basis for a complex inner product space and

$$z = c_1 u_1 + \cdots + c_n u_n$$

then

$$c_j = \langle z, u_j \rangle \quad \text{and} \quad \overline{c_j} = \langle u_j, z \rangle$$

5. There will not be a unique unitary diagonalizing matrix for a given Hermitian matrix A, however, the column vectors of any unitary diagonalizing matrix must be unit eigenvectors of A.

(g) **Solution.** The matrix has rank 1, so its nullity must be 2. Therefore 0 is an eigenvalue whose multiplicity is at least 2. If we take $\lambda_1 = \lambda_2 = 0$, then since the trace of the matrix is 6, λ_3 must equal 6. The eigenvectors belonging to $\lambda_3 = 6$ are nonzero multiples of $x_3 = (-2, \ -1, \ 1)^T$. To obtain a unit eigenvector we set

$$q_3 = \frac{1}{\|x_1\|} x_1 = \frac{1}{\sqrt{6}} \begin{pmatrix} -2 \\ -1 \\ 1 \end{pmatrix}$$

The vectors $x_1 = (1, \ 0, \ 2)^T$ and $x_2 = (-1, \ 2, \ 0)^T$ form a basis for the eigenspace corresponding to $\lambda = 0$, however, they do not form an orthonormal basis. To obtain the orthonormal basis we must use the Gram–Schmidt process.

$$r_{11} = \|x_1\| = \sqrt{5}$$
$$q_1 = \frac{1}{\sqrt{5}} x_1 = \frac{1}{\sqrt{5}}(1, \ 0, \ 2)^T$$
$$p_1 = (x_2^T q_1)q_1 = -\frac{1}{\sqrt{5}} q_1 = -\frac{1}{5}(1, \ 0, \ 2)^T$$
$$x_2 - p_1 = \left(-\frac{4}{5}, \ 2, \ \frac{2}{5}\right)^T$$

$$r_{22} = \|\mathbf{x}_2 - \mathbf{p}_1\| = \frac{2\sqrt{30}}{5}$$

$$\mathbf{q}_2 = \frac{1}{\sqrt{30}}(-2, \ 5, \ 1)^T$$

The eigenvectors $\mathbf{q}_1, \mathbf{q}_2, \mathbf{q}_3$ form an orthonormal set. Therefore

$$Q = (\mathbf{q}_1, \mathbf{q}_2, \mathbf{q}_3) = \begin{pmatrix} \frac{1}{\sqrt{5}} & -\frac{2}{\sqrt{30}} & -\frac{2}{\sqrt{6}} \\ 0 & \frac{5}{\sqrt{30}} & -\frac{1}{\sqrt{6}} \\ \frac{2}{\sqrt{5}} & \frac{1}{\sqrt{30}} & \frac{1}{\sqrt{6}} \end{pmatrix}$$

is an orthogonal diagonalizing matrix.

7. (c) **Hint.** Use the result from Exercise 17(b) of Section 1 of this chapter.

9. **Hint.** $< \mathbf{z}, \alpha\mathbf{x} + \beta\mathbf{y} > = \overline{< \alpha\mathbf{x} + \beta\mathbf{y}, \mathbf{z} >}$

10. **Hint.** Show first that

$$< \mathbf{z}, \mathbf{w} > = \sum_{i=1}^{n} a_i < \mathbf{u}_i, \mathbf{w} >$$

11. **Hint.** Factor A into a product $A = QDQ^H$ and find a diagonal matrix D_1 so that $D_1^H D_1 = D$.

14. **Hint.** If U is a matrix that is both unitary and Hermitian, then

$$U^2 = U^H U = I$$

15. (b) and (c) **Hint.** If A has Schur decomposition UTU^H, then $AU = UT$. So AU and UT have the same column vectors.

16. **Hint.** See Exercise 19 in the previous section of the book.

18. **Hint.** The proof is similar to the proof that the eigenvalues of an Hermitian matrix must all be real.

20. **Hint.** The matrix B is similar to A and it should turn out to be symmetric.

21. (b) **Hint.** Let $B = A^{-1}CA$. The matrices B and C are similar and the eigenvalues of C are the roots of $p(x)$.

22. **Hint.** If A is Hermitian, then there is a unitary U that diagonalizes A and we can factor A into a product UDU^H where D is a diagonal matrix.

5 | SINGULAR VALUE DECOMPOSITION

Overview

In this section we present one of the most important matrix factorizations, the singular value decomposition. The factorization takes the form $A = U\Sigma V^T$, where

U and V are orthogonal matrices and Σ is an $m \times n$ diagonal matrix. The diagonal entries of Σ are the singular values of A. The column vectors of U and V are the singular vectors of A. They can be used to form orthonormal bases for each of the four fundamental subspaces associated with A. The singular values tell us how close the matrix is to lower rank matrices. When matrix computations are performed by computer, the singular values allow us to find the effective rank of a matrix relative to the precision of the machine.

Important Concepts

1. **Singular Value Decomposition.** The *singular value decomposition* of an $m \times n$ matrix A is a factorization of the form $A = U\Sigma V^T$, where U is an $m \times m$ orthogonal matrix, V is an $n \times n$ orthogonal matrix, and Σ is an $m \times n$ matrix whose off-diagonal entries are all 0's and whose diagonal elements satisfy

$$\sigma_1 \geq \sigma_2 \geq \cdots \geq \sigma_n \geq 0$$

Thus Σ is a matrix of the form

$$\Sigma = \begin{pmatrix} \sigma_1 & & & & \\ & \sigma_2 & & & \\ & & \ddots & & \\ & & & \sigma_n & \\ & & & & \end{pmatrix}$$

The σ_i's determined by this factorization are unique and are called the *singular values* of A.

2. **Singular vectors.** Let A be a matrix with singular value decomposition $U\Sigma V^T$. The column vectors of U and V are the *singular vectors* of A. The $\mathbf{v}_j$'s are called the *right singular vectors* of A and the $\mathbf{u}_j$'s are called the *left singular vectors* of A.

3. **Numerical rank.** The *numerical rank* of an $m \times n$ matrix is the number of singular values of the matrix that are greater than $\sigma_1 \max(m, n)\epsilon$, where σ_1 is the largest singular value of A and ϵ is the machine epsilon.

Important Theorems and Results

▶ THEOREM 6.5.1 **(The SVD Theorem)** *If A is an $m \times n$ matrix, then A has a singular value decomposition.*

▶ The rank of a matrix A is equal to the number of its nonzero singular values (where singular values are counted according to multiplicity).

▶ If A is a $m \times n$ matrix of rank r with singular value decomposition $U\Sigma V^T$, then

(i) $\mathbf{v}_1, \ldots, \mathbf{v}_r$ form an orthonormal basis for $R(A^T)$.

(ii) $\mathbf{v}_{r+1}, \ldots, \mathbf{v}_n$ form an orthonormal basis for $N(A)$.

(iii) $\mathbf{u}_1, \ldots, \mathbf{u}_r$ form an orthonormal basis for $R(A)$.

(iv) $\mathbf{u}_{r+1}, \ldots, \mathbf{u}_m$ form an orthonormal basis for $N(A^T)$.

▶ THEOREM 6.5.2. *Let $A = U\Sigma V^T$ be an $m \times n$ matrix and let $\mathcal{M}$ denote the set of all $m \times n$ matrices of rank k or less, where $0 < k < \operatorname{rank}(A)$. If X is a matrix in $\mathcal{M}$ satisfying*

$$\|A - X\|_F = \min_{S \in \mathcal{M}} \|A - S\|_F$$

then

$$\|A - X\|_F = \left(\sigma_{k+1}^2 + \sigma_{k+2}^2 + \cdots + \sigma_n^2\right)^{1/2}$$

In particular, if $A' = U\Sigma'V^T$, where

$$\Sigma' = \begin{pmatrix} \sigma_1 & & & \\ & \ddots & & O \\ & & \sigma_k & \\ \hline & O & & O \end{pmatrix} = \begin{pmatrix} \Sigma_k & O \\ O & O \end{pmatrix}$$

then

$$\|A - A'\|_F = \left(\sigma_{k+1}^2 + \cdots + \sigma_n^2\right)^{1/2} = \min_{S \in \mathcal{M}} \|A - S\|_F$$

▶ If A has singular value decomposition $U\Sigma V^T$, then A can be expressed as an outer product expansion of its left and right singular vectors.

$$A = \sigma_1 \mathbf{u}_1 \mathbf{v}_1^T + \sigma_2 \mathbf{u}_2 \mathbf{v}_2^T + \cdots + \sigma_n \mathbf{u}_n \mathbf{v}_n^T$$

▶ If A is an $m \times n$ matrix with singular values $\sigma_1, \sigma_2, \ldots, \sigma_n$, then

$$\|A\|_F = \left(\sigma_1^2 + \sigma_2^2 + \cdots + \sigma_n^2\right)^{1/2}$$

Exercises: Solutions and Hints

2. (d) **Solution.** To start we find the eigenvalues and eigenvectors of $A^T A$.

$$A^T A = \begin{pmatrix} 2 & 0 & 0 & 0 \\ 0 & 2 & 1 & 0 \\ 0 & 1 & 2 & 0 \end{pmatrix} \begin{pmatrix} 2 & 0 & 0 \\ 0 & 2 & 1 \\ 0 & 1 & 2 \\ 0 & 0 & 0 \end{pmatrix} = \begin{pmatrix} 4 & 0 & 0 \\ 0 & 5 & 4 \\ 0 & 4 & 5 \end{pmatrix}$$

The eigenvalues of $A^T A$ are $\lambda_1 = 9$, $\lambda_2 = 4$ and $\lambda_3 = 1$. The singular values are square roots of the eigenvalues.

$$\sigma_1 = 3, \quad \sigma_2 = 2, \quad \sigma_3 = 1$$

The eigenvectors of $A^T A$ are

$$\mathbf{x}_1 = \begin{pmatrix} 0 \\ 1 \\ 1 \end{pmatrix}, \quad \mathbf{x}_2 = \begin{pmatrix} 1 \\ 0 \\ 0 \end{pmatrix}, \quad \mathbf{x}_3 = \begin{pmatrix} 0 \\ -1 \\ 1 \end{pmatrix},$$

The eigenvectors are mutually orthogonal, so to obtain an orthonormal set we need only normalize,

$$\mathbf{v}_1 = \frac{1}{\sqrt{2}} \mathbf{x}_1 = \begin{pmatrix} 0 \\ \frac{1}{\sqrt{2}} \\ \frac{1}{\sqrt{2}} \end{pmatrix}, \quad \mathbf{v}_2 = \mathbf{x}_2 = \begin{pmatrix} 1 \\ 0 \\ 0 \end{pmatrix}, \quad \mathbf{v}_3 = \frac{1}{\sqrt{2}} \mathbf{x}_3 = \begin{pmatrix} 0 \\ -\frac{1}{\sqrt{2}} \\ \frac{1}{\sqrt{2}} \end{pmatrix}$$

For $j = 1, 2, 3$ the left singular vectors are determined by setting $\mathbf{u}_j = \frac{1}{\sigma_j} \mathbf{v}_j$. Thus

$$\mathbf{u}_1 = \frac{1}{3} A \mathbf{v}_1 = \frac{1}{3} \begin{pmatrix} 2 & 0 & 0 \\ 0 & 2 & 1 \\ 0 & 1 & 2 \\ 0 & 0 & 0 \end{pmatrix} \begin{pmatrix} 0 \\ \frac{1}{\sqrt{2}} \\ \frac{1}{\sqrt{2}} \end{pmatrix} = \begin{pmatrix} 0 \\ \frac{1}{\sqrt{2}} \\ \frac{1}{\sqrt{2}} \\ 0 \end{pmatrix}$$

The second left singular vector is

$$\mathbf{u}_2 = \frac{1}{2} A \mathbf{v}_2 = \frac{1}{2} \mathbf{a}_1 = \begin{pmatrix} 1 \\ 0 \\ 0 \\ 0 \end{pmatrix}$$

The third left singular vector is

$$\mathbf{u}_3 = A \mathbf{v}_3 = \begin{pmatrix} 0 \\ -\frac{1}{\sqrt{2}} \\ \frac{1}{\sqrt{2}} \\ 0 \end{pmatrix}$$

The remaining left singular vector $\mathbf{u}_4$ must come from $N(A^T)$. All solutions to $A^T\mathbf{x} = \mathbf{0}$ are multiples of $(0, 0, 0, 1)^T$. Thus $\mathbf{u}_4 = (0, 0, 0, 1)^T$. The singular value decomposition of A is

$$A = U\Sigma V^T = \begin{pmatrix} 0 & 1 & 0 & 0 \\ \frac{1}{\sqrt{2}} & 0 & -\frac{1}{\sqrt{2}} & 0 \\ -\frac{1}{\sqrt{2}} & 0 & \frac{1}{\sqrt{2}} & 0 \\ 0 & 0 & 0 & 1 \end{pmatrix} \begin{pmatrix} 3 & 0 & 0 \\ 0 & 2 & 0 \\ 0 & 0 & 1 \\ 0 & 0 & 0 \end{pmatrix} \begin{pmatrix} 0 & \frac{1}{\sqrt{2}} & \frac{1}{\sqrt{2}} \\ 1 & 0 & 0 \\ 0 & -\frac{1}{\sqrt{2}} & \frac{1}{\sqrt{2}} \end{pmatrix}$$

6. **Hint.** $A^T A = A^2$.

8. **Hint.** $A^T A = V\Sigma^T \Sigma V^T$ and $AA^T = U\Sigma\Sigma^T U^T$.

9. **Solution.** If σ is a singular value of A, then σ^2 is an eigenvalue of $A^T A$. Let $\mathbf{x}$ be an eigenvector of $A^T A$ belonging to σ^2. It follows that

$$A^T A\mathbf{x} = \sigma^2\mathbf{x}$$
$$\mathbf{x}^T A^T A\mathbf{x} = \sigma^2\mathbf{x}^T\mathbf{x}$$
$$\|A\mathbf{x}\|_2^2 = \sigma^2\|\mathbf{x}\|_2^2$$
$$\sigma = \frac{\|A\mathbf{x}\|_2}{\|\mathbf{x}\|_2}$$

10. **Hint.** Plug

$$\hat{\mathbf{x}} = A^+\mathbf{b} = U\Sigma^+ V^T\mathbf{b}$$

into the normal equations.

6 | QUADRATIC FORMS

Overview

Up until now we have focused on the role that matrices play in the study of linear equations. In this section we will see that matrices also play an important role in the study of quadratic equations. With each quadratic equation we can associate a vector function $f(\mathbf{x}) = \mathbf{x}^T A\mathbf{x}$. Such a vector function is called a *quadratic form*. Quadratic forms arise in a wide variety of applied problems. They are particularly important in the study of optimization theory.

Important Concepts

1. **Quadratic forms.** If A is a symmetric $n \times n$ matrix, then the vector function $f(\mathbf{x}) = \mathbf{x}^T A\mathbf{x}$ is said to be a *quadratic form* in the variables $x_1, x_2, \ldots, x_n$.

2. **Stationary point.** Let $F(\mathbf{x})$ be a real-valued vector function on R^n. A point $\mathbf{x}_0$ in R^n is said to be a *stationary point* of F if all of the first partial derivatives of F at $\mathbf{x}_0$ exist and are equal to 0.

3. **Positive and negative definite quadratic forms.** A quadratic form $f(\mathbf{x}) = \mathbf{x}^T A \mathbf{x}$ is said to be *definite* if it takes on only one sign as $\mathbf{x}$ varies over all nonzero vectors in R^n. The form is *positive definite* if $\mathbf{x}^T A \mathbf{x} > 0$ for all nonzero $\mathbf{x}$ in R^n and *negative definite* if $\mathbf{x}^T A \mathbf{x} < 0$ for all nonzero $\mathbf{x}$ in R^n. A quadratic form is said to be *indefinite* if it takes on values that differ in sign. If $f(\mathbf{x}) = \mathbf{x}^T A \mathbf{x} \geq 0$ and assumes the value 0 for some $\mathbf{x} \neq \mathbf{0}$, then $f(\mathbf{x})$ is said to be *positive semidefinite*. If $f(\mathbf{x}) \leq 0$ and assumes the value 0 for some $\mathbf{x} \neq \mathbf{0}$, then $f(\mathbf{x})$ is said to be *negative semidefinite*.

4. **Positive and negative definite matrices.** A real symmetric matrix A is said to be
   (i) *Positive definite* if $\mathbf{x}^T A \mathbf{x} > 0$ for all nonzero $\mathbf{x}$ in R^n
   (ii) *Negative definite* if $\mathbf{x}^T A \mathbf{x} < 0$ for all nonzero $\mathbf{x}$ in R^n
   (iii) *Positive semidefinite* if $\mathbf{x}^T A \mathbf{x} \geq 0$ for all nonzero $\mathbf{x}$ in R^n
   (iv) *Negative semidefinite* if $\mathbf{x}^T A \mathbf{x} \leq 0$ for all nonzero $\mathbf{x}$ in R^n

5. **Hessian.** Let $F(\mathbf{x}) = F(x_1, \ldots, x_n)$ be a real-valued function whose third partial derivatives are all continuous. Let $\mathbf{x}_0$ be a fixed vector in R^n and define the matrix $H = H(\mathbf{x}_0)$ by

$$h_{ij} = F_{x_i x_j}(\mathbf{x}_0)$$

$H(\mathbf{x}_0)$ is called the *Hessian* of F at $\mathbf{x}_0$.

Important Theorems and Results

▶ THEOREM 6.6.1 **(Principal Axes Theorem)** *If A is a real symmetric $n \times n$ matrix, then there is a change of variables $\mathbf{u} = Q^T \mathbf{x}$ such that $\mathbf{x}^T A \mathbf{x} = \mathbf{u}^T D \mathbf{u}$, where D is a diagonal matrix.*

▶ THEOREM 6.6.2. *Let A be a real symmetric $n \times n$ matrix. Then A is positive definite if and only if all its eigenvalues are positive.*

▶ If $\mathbf{x}_0$ is a stationary point of a vector function $F(\mathbf{x})$ and $H(\mathbf{x}_0)$ is the Hessian of F at $\mathbf{x}_0$, then we can classify the stationary point as follows:

   (i) $\mathbf{x}_0$ is a local minimum of F if $H(\mathbf{x}_0)$ is positive definite.
   (ii) $\mathbf{x}_0$ is a local maximum of F if $H(\mathbf{x}_0)$ is negative definite.
   (iii) $\mathbf{x}_0$ is a saddle point of F if $H(\mathbf{x}_0)$ is indefinite.

Exercises: Solutions and Hints

3. (b) **Solution.** The matrix for the quadratic form is

$$A = \begin{pmatrix} 3 & 4 \\ 4 & 3 \end{pmatrix}$$

The eigenvalues of A are $\lambda_1 = 7$, $\lambda_2 = -1$. Since A is symmetric we can find a pair of orthonormal eigenvectors

$$\mathbf{q}_1 = \begin{pmatrix} \frac{1}{\sqrt{2}} \\ \frac{1}{\sqrt{2}} \end{pmatrix} \qquad \text{and} \qquad \mathbf{q}_1 = \begin{pmatrix} -\frac{1}{\sqrt{2}} \\ \frac{1}{\sqrt{2}} \end{pmatrix}$$

Let

$$Q = \frac{1}{\sqrt{2}} \begin{pmatrix} 1 & -1 \\ 1 & 1 \end{pmatrix} \qquad \text{and} \qquad \begin{pmatrix} x' \\ y' \end{pmatrix} = Q^T \begin{pmatrix} x \\ y \end{pmatrix}$$

With this change of variables the equation simplifies to

$$7(x')^2 - (y')^2 = -28$$

This is the equation of an hyperbola in the $x'y'$ coordinate system. We can rewrite the equation in standard form

$$\frac{(y')^2}{28} - \frac{(x')^2}{4} = 1$$

6. (f) **Solution.** The eigenvalues are $\lambda_1 = 8$, $\lambda_2 = 2$, $\lambda_3 = 2$. Since all of the eigenvalues are positive, the matrix is positive definite.

7. (d) **Solution.** The first partials of f are

$$f_x = -\frac{2y}{x^3} + \frac{1}{y^2} + y$$

$$f_y = \frac{1}{x^2} - \frac{2x}{y^3} + x$$

and the second partials are

$$f_{xx} = \frac{6y}{x^4}$$

$$f_{xy} = f_{yx} = -\frac{2}{x^3} - \frac{2}{y^3} + 1$$

$$f_{yy} = \frac{6x}{y^4}$$

Setting $x = y = 1$ we see that the Hessian of f at $(1, 1)$ is

$$H = \begin{pmatrix} 6 & -3 \\ -3 & 6 \end{pmatrix}$$

The eigenvalues of H are $\lambda_1 = 9$, $\lambda_2 = 3$. Since both are positive, the matrix is positive definite and hence $(1, 1)$ is a local minimum.

8. **Hint.** Can you find a 2×2 symmetric matrix A with positive determinant for which $\mathbf{e}_1^T A \mathbf{e}_1 < 0$?

9. **Hint.** If A is symmetric positive definite, then what can you conclude about its eigenvalues?

10. **Hint.** Make use of Theorem 1.4.2 to show $A^T A$ is not positive definite.

12. **Hint.** See the hint for Exercise 8.

13. **Hint.** If $\mathbf{x}$ is a nonzero vector and S is nonsingular, then $\mathbf{y} = S\mathbf{x}$ must also be a nonzero vector.

7   POSITIVE DEFINITE MATRICES

Overview

Symmetric positive definite matrices arise in a wide variety of applications. In this section we study some of the main properties of these matrices.

Important Concepts

1. **Positive definite matrices.** A symmetric $n \times n$ matrix A is *positive definite* if $\mathbf{x}^T A \mathbf{x} > 0$ for all nonzero vectors $\mathbf{x}$ in R^n.

2. **Leading principal submatrices.** Given an $n \times n$ matrix A, let A_r denote the matrix formed by deleting the last $n - r$ rows and columns of A. A_r is called the *leading principal submatrix* of A of order r.

3. **Cholesky factorization.** If A is symmetric positive definite, then A can be factored into product LL^T where L is lower triangular. This decomposition is known as the *Cholesky factorization*. Note, this factorization is often written in the form $A = R^T R$ where $R = L^T$ is upper triangular.

Important Theorems

▶ THEOREM 6.7.1. *Let A be a symmetric $n \times n$ matrix. The following are equivalent.*

(a) *A is positive definite.*

(b) *The leading principal submatrices $A_1, \ldots, A_n$ all have positive determinants.*

(c) *A can be reduced to upper triangular form using only row operation III and the pivot elements will all be positive.*

(d) *A has a Cholesky factorization LL^T (where L is lower triangular with positive diagonal entries).*

(e) *A can be factored into a product $B^T B$ for some nonsingular matrix B.*

Exercises: Solutions and Hints

4. (d) **Solution.** First we compute a triangular factorization of the matrix using only row operation III. At each step we keep track of the multiples of

the pivot row that are subtracted from the other rows.

$$
\begin{pmatrix} 9 & 3 & -6 \\ 3 & 4 & 1 \\ -6 & 1 & 9 \end{pmatrix} \rightarrow \begin{pmatrix} 9 & 3 & -6 \\ 0 & 3 & 3 \\ 0 & 3 & 5 \end{pmatrix} \begin{array}{l} \\ l_{21} = \frac{1}{3} \\ l_{31} = -\frac{2}{3} \end{array}
$$

$$
\rightarrow \begin{pmatrix} 9 & 3 & -6 \\ 0 & 3 & 3 \\ 0 & 0 & 2 \end{pmatrix} \begin{array}{l} \\ \\ l_{32} = 1 \end{array}
$$

The triangular factorization is

$$
\begin{pmatrix} 9 & 3 & -6 \\ 3 & 4 & 1 \\ -6 & 1 & 9 \end{pmatrix} = \begin{pmatrix} 1 & 0 & 0 \\ \frac{1}{3} & 1 & 0 \\ -\frac{2}{3} & 1 & 1 \end{pmatrix} \begin{pmatrix} 9 & 3 & -6 \\ 0 & 3 & 3 \\ 0 & 0 & 2 \end{pmatrix}
$$

If we express the upper triangular U as the product of a diagonal matrix times a unit upper triangular matrix, then we have the desired LDL^T factorization.

$$
\begin{pmatrix} 9 & 3 & -6 \\ 3 & 4 & 1 \\ -6 & 1 & 9 \end{pmatrix} = \begin{pmatrix} 1 & 0 & 0 \\ \frac{1}{3} & 1 & 0 \\ -\frac{2}{3} & 1 & 1 \end{pmatrix} \begin{pmatrix} 9 & 0 & 0 \\ 0 & 3 & 0 \\ 0 & 0 & 2 \end{pmatrix} \begin{pmatrix} 1 & \frac{1}{3} & -\frac{2}{3} \\ 0 & 1 & 1 \\ 0 & 0 & 1 \end{pmatrix}
$$

5. (d) **Solution.** To obtain the Cholesky factor L we right multiply the unit lower triangular matrix by $D^{1/2}$.

$$
L = \begin{pmatrix} 1 & 0 & 0 \\ \frac{1}{3} & 1 & 0 \\ -\frac{2}{3} & 1 & 1 \end{pmatrix} \begin{pmatrix} 3 & 0 & 0 \\ 0 & \sqrt{3} & 0 \\ 0 & 0 & \sqrt{2} \end{pmatrix} = \begin{pmatrix} 3 & 0 & 0 \\ 1 & \sqrt{3} & 0 \\ -2 & \sqrt{3} & \sqrt{2} \end{pmatrix}
$$

Our original matrix will factor into a product LL^T.

$$
\begin{pmatrix} 9 & 3 & -6 \\ 3 & 4 & 1 \\ -6 & 1 & 9 \end{pmatrix} = \begin{pmatrix} 3 & 0 & 0 \\ 1 & \sqrt{3} & 0 \\ -2 & \sqrt{3} & \sqrt{2} \end{pmatrix} \begin{pmatrix} 3 & 1 & -2 \\ 0 & \sqrt{3} & \sqrt{3} \\ 0 & 0 & \sqrt{2} \end{pmatrix}
$$

6. **Hint.** You must show that the three conditions in the definition of an inner product are all satisfied.

9. **Hint.** Show first that if B is an $m \times n$ matrix of rank n and $\mathbf{x}$ is a nonzero vector in R^n, then $\mathbf{y} = B\mathbf{x}$ is a nonzero vector in R^m.

10. **Hint.** Use the definition of the matrix exponential to show $(e^A)^T = e^A$. How are the eigenvalues of A and e^A related?

11. **Hint.** If B is symmetric then $B^2 = B^T B$.

12. (b) **Hint.** To show B^2 is not positive definite you must find a nonzero vector $\mathbf{x}$ such that $\mathbf{x}^T B^2 \mathbf{x} \leq 0$.

13. (b) **Hint.** The proof is similar to the proof that the leading principal submatrices of a positive definite matrix are all positive definite.

14. (b) **Solution.**
    Cholesky Factorization Algorithm
    Set $L_1 = (\sqrt{a_{11}})$
    For $k = 1, \ldots, n-1$
        (1) Let $\mathbf{y}_k$ be the vector consisting of the first k entries of $\mathbf{a}_{k+1}$ and let β_k be the $(k+1)$st entry of $\mathbf{a}_{k+1}$.
        (2) Solve the lower triangular system $L_k \mathbf{x}_k = \mathbf{y}_k$ for $\mathbf{x}_k$.
        (3) Set $\alpha_k = (\beta_k - \mathbf{x}_k^T \mathbf{x}_k)^{1/2}$
        (4) Set

$$L_{k+1} = \begin{pmatrix} L_k & 0 \\ \mathbf{x}_k^T & \alpha_k \end{pmatrix}$$

End (For Loop)
$L = L_n$

The Cholesky decomposition of A is LL^T.

8 | NONNEGATIVE MATRICES

Overview

In many of the types of linear systems that occur in applications, the entries of the coefficient matrix represent nonnegative quantities. This section deals with the study of such matrices and some of their properties.

Important Concepts

1. **Nonnegative and positive matrices.** An $n \times n$ matrix A with real entries is said to be *nonnegative* if $a_{ij} \geq 0$ for each i and j and *positive* if $a_{ij} > 0$ for each i and j. Similarly, a vector $\mathbf{x} = (x_1, \ldots, x_n)^T$ is said to be *nonnegative* if each $x_i \geq 0$ and *positive* if each $x_i > 0$.

2. **Reducible and irreducible matrices.** A nonnegative matrix A is said to be *reducible* if there exists a partition of the index set $\{1, 2, \ldots, n\}$ into

nonempty disjoint sets I_1 and I_2 such that $a_{ij} = 0$ whenever $i \in I_1$ and $j \in I_2$. Otherwise, A is said to be *irreducible*.

3. **Regular Markov Process.** A Markov process is *regular* if all of the entries of some power of the transition matrix are strictly positive.

Important Theorems and Results

▶ THEOREM 6.8.1 **(Perron)** *If A is a positive $n \times n$ matrix, A has a positive real eigenvalue r with the following properties:*

(i) *r is a simple root of the characteristic equation.*

(ii) *r has a positive eigenvector $\mathbf{x}$.*

(iii) *If λ is any other eigenvalue of A, then $|\lambda| < r$.*

▶ THEOREM 6.8.2 **(Frobenius)** *If A is an irreducible nonnegative matrix, then A has a positive real eigenvalue r with the following properties:*

(i) *r has a positive eigenvector $\mathbf{x}$.*

(ii) *If λ is any other eigenvalue of A, then $|\lambda| \leq r$. The eigenvalues of modulus r are all simple roots of the characteristic equation. Indeed, if there are m eigenvalues of modulus r, they must be of the form*

$$\lambda_k = re^{\frac{2k\pi i}{m}} \qquad k = 0, 1, \ldots, m - 1$$

▶ If a Markov process is regular then it has a unique steady state vector and for any starting probability vector $\mathbf{x}_0$, the Markov chain will converge to the steady state vector.

Exercises: Solutions and Hints

7. (b) **Solution.** If we partition the index set $\{1, 2, 3, 4\}$ into two disjoint sets

$$I_1 = \{1, 3, 4\} \qquad \text{and} \qquad I_2 = \{2\}$$

then the only entries a_{ij} for which both $i \in I_1$ and $j \in I_2$ are a_{12}, a_{32}, a_{42} and all of these entries are equal to 0. Therefore the matrix is reducible. The permutation matrix P that is used to transform A into the desired block form can be constructed as a product of elementary matrices of type I. Since we are operating on both sides of the matrix, whenever we interchange two rows, say rows k and l, then we must also interchange columns k and l. The sequence of row swaps that we use in the transformation of A will, when applied to the identity matrix, transform it to the desired permutation matrix P. Let E_{ij} denote the elementary matrix formed from I by interchanging its ith and jth rows. We proceed as follows. At the first step we interchange rows

2 and 3 of A and then interchange columns 2 and 3 of the resulting matrix. This is equivalent to multiplying A on both sides by E_{23}.

$$E_{23}AE_{23} = \begin{bmatrix} 1 & 1 & 0 & 1 \\ 1 & 1 & 0 & 1 \\ 1 & 1 & 1 & 1 \\ 1 & 1 & 0 & 1 \end{bmatrix} \qquad E_{23}I = \begin{bmatrix} 1 & 0 & 0 & 0 \\ 0 & 0 & 1 & 0 \\ 0 & 1 & 0 & 0 \\ 0 & 0 & 0 & 1 \end{bmatrix}$$

Next we interchange rows 3 and 4 and then columns 3 and 4.

$$E_{34}E_{23}AE_{23}E_{34} = \begin{bmatrix} 1 & 1 & 1 & 0 \\ 1 & 1 & 1 & 0 \\ 1 & 1 & 1 & 0 \\ 1 & 1 & 1 & 1 \end{bmatrix} \qquad E_{34}E_{23}I = \begin{bmatrix} 1 & 0 & 0 & 0 \\ 0 & 0 & 1 & 0 \\ 0 & 0 & 0 & 1 \\ 0 & 1 & 0 & 0 \end{bmatrix}$$

If we set

$$P = E_{34}E_{23}I = E_{34}E_{23}$$

then

$$P^T = E_{23}^T E_{34}^T = E_{23}E_{34}$$

and

$$PAP^T = E_{34}E_{23}AE_{23}E_{34}$$

has the desired block structure.

8. **Hint.** See Theorem 6.7.2.

9. (b) **Hint.** Apply Perron's theorem to B and C.

11. **Hint.** Make use of the result from Exercise 10.

12. **Hint.** Apply Perron's theorem to A^k.

13. (c) **Hint.** Show that if c_1 did equal 0, then $\mathbf{y}_j$ would approach the zero vector as $j \to \infty$, contradicting the result from part (b).

14. **Solution.** In general if the matrix is nonnegative then there is no guarantee that it has a dominant eigenvalue with a positive eigenvector. So the results from parts (c) and (d) of Exercise 13 would not hold in this case. On the other hand if A^k is a positive matrix for some k, then by Exercise 12, $\lambda_1 = 1$ is the dominant eigenvalue of A and it has a positive eigenvector $\mathbf{x}_1$. Therefore the results from Exercise 13 will be valid in this case.

MATLAB EXERCISES

1. Initially $\mathbf{x} = \mathbf{e}_1$, the standard basis vector, and

$$A\mathbf{x} = \frac{5}{4}\mathbf{e}_1 = \frac{5}{4}\mathbf{x}$$

is in the same direction as $\mathbf{x}$. So $\mathbf{x}_1 = \mathbf{e}_1$ is an eigenvector of A belonging to the eigenvalue $\lambda_1 = \frac{5}{4}$. When the initial vector is rotated so that $\mathbf{x} = \mathbf{e}_2$ the image will be

$$Ax = \frac{3}{4}\mathbf{e}_2 = \frac{3}{4}\mathbf{x}$$

so $\mathbf{x}_2 = \mathbf{e}_2$ is an eigenvector of A belonging to the eigenvalue $\lambda_2 = \frac{3}{4}$. The second diagonal matrix has the same first eigenvalue-eigenvector pair and the second eigenvector is again $\mathbf{x}_2 = \mathbf{e}_2$, however, this time the eigenvalue is negative since $\mathbf{x}_2$ and $A\mathbf{x}_2$ are in opposite directions. In general for any 2×2 diagonal matrix D, the eigenvalues will be d_{11} and d_{22} and the corresponding eigenvectors will be $\mathbf{e}_1$ and $\mathbf{e}_2$.

2. For the identity matrix the eigenvalues are the diagonal entries so $\lambda_1 = \lambda_2 = 1$. In this case not only are $\mathbf{e}_1$ and $\mathbf{e}_2$ eigenvectors, but any vector $\mathbf{x} = x_1\mathbf{e}_1 + x_2\mathbf{e}_2$ is an eigenvector.

7. The tenth matrix is singular, so one of its eigenvalues is 0. To find the eigenvector using the `eigshow` utility you most rotate $\mathbf{x}$ until $A\mathbf{x}$ coincides with the zero vector. The other eigenvalue of this matrix is $\lambda_2 = 1.5$. Since the eigenvalues are distinct their corresponding eigenvectors must be linearly independent. The next two matrices both have multiple eigenvalues and both are defective. Thus for either matrix any pair of eigenvectors would be linearly dependent.

8. The characteristic polynomial of a 2×2 matrix is a quadratic polynomial and its graph will be a parabola. The eigenvalues will be equal when the graph of the parabola corresponding to the characteristic polynomial has its vertex on the x axis. For a random 2×2 matrix the probability that this will happen should be 0.

11. (a) **Hint.** What is the nullity of $A - I$? How is the multiplicity of the eigenvalue related to the dimension of the eigenspace? For symmetric matrices, eigenvalue computations should be quite accurate. Thus one would expect to get nearly full machine accuracy in the computed eigenvalues of A even though $\lambda = 1$ is a multiple eigenvalue.

    (b) The roots of a tenth degree polynomial are quite sensitive, i.e., any small roundoff errors in either the data or in the computations are liable to lead to significant errors in the computed roots. In particular if $p(\lambda)$ has multiple roots, the computed eigenvalues are label to be complex.

13. **Hint.** Apply the Rank-Nullity theorem.

14. **Hint.** See Exercise 7 in Section 1 of this chapter.

15. In theory A and B should have the same eigenvalues. However for a defective matrix it is difficult to compute the eigenvalues accurately. Thus even though B would be defective if computed in exact arithmetic, the matrix computed using floating point arithmetic may have distinct eigenvalues and the computed matrix X of eigenvectors may turn out to be nonsingular. If, however, `rcond` is very small, this would indicate that the column vectors of X are nearly dependent and hence that B may be defective.

18. (a) By construction S has integer entries and $\det(S) = 1$. It follows that $S^{-1} = \operatorname{adj} S$ will also have integer entries. (See Section 3 of Chapter 2.)

21. If the singular values of A are s_1, s_2 and $\mathbf{u}_1$ is the left singular vector corresponding to $\mathbf{x}$, then $A\mathbf{x} = s_1\mathbf{u}_1$. Thus the image $A\mathbf{x}$ is a vector in the direction of $\mathbf{u}_1$ with length s_1. If $\mathbf{u}_2$ is the left singular vector corresponding to $\mathbf{y}$ then $A\mathbf{y} = s_2\mathbf{u}_2$. Thus the image $A\mathbf{y}$ is a vector in the direction of $\mathbf{u}_2$ with length s_2.

If you rotate the axes a full $360°$ the image vectors will trace out an ellipse. The major axis of the ellipse will be the line corresponding to the span of $\mathbf{u}_1$ and the diameter of the ellipse along its major axis will be $2s_1$ The minor axis of the ellipse will be the line corresponding to the span of $\mathbf{u}_2$ and the diameter of the ellipse along its minor axis will be $2s_2$.

22. The stationary points of the Hessian are $(-\frac{1}{4}, 0)$ and $-\frac{71}{4}, 4)$. If the stationary values are substituted into the Hessian, then in each case we can compute the eigenvalues using the MATLAB's `eig` command. If we use the `double` command to view the eigenvalues in numeric format, the displayed values should be 7.6041 and -2.1041 for the first stationary point and -7.6041, 2.1041 for the second stationary point. Thus both stationary points are saddle points.

25. (a) If you subtract 1 from the $(6,6)$ entry of P, the resulting matrix will be singular.

(c) The matrix P is symmetric. The leading principal submatrices of P are all Pascal matrices. If all have determinant equal to 1, then all have positive determinants. Therefore P should be positive definite. The Cholesky factor R is a unit upper triangular matrix. Therefore

$$\det(P) = \det(R^T)\det(R) = 1$$

(d) If one sets $r_{88} = 0$, then R becomes singular. It follows that Q must also be singular since

$$\det(Q) = \det(R^T)\det(R) = 0$$

Since R is upper triangular, when one sets $r_{88} = 0$ it will only affect the $(8,8)$ entry of the product $R^T R$. Since R has 1's on the diagonal, changing r_{88} from 1 to 0 will have the effect of decreasing the $(8,8)$ entry of $R^T R$ by 1.

CHAPTER TEST A

1. Hint. If the eigenvalues of A are all nonzero then what can you conclude about the value of $\det(A)$?

3. Hint. Whether or not the matrix is defective depends upon the number of linear independent eigenvectors.

4. Hint. The eigenspace corresponding to $\lambda = 0$ is $N(A)$, so the dimension of the eigenspace is equal to the nullity of A.

5. **Hint.** The hint for question 4 also applies to this question.
6. **Hint.** Look at some triangular matrices that have some 0's on the diagonal.
7. **Hint.** See the list of observations following the proof of Theorem 6.5.1
8. **Hint.** $U^{-1} = U^H$, so $A = UTU^{-1}$.
9. **Hint.** What do we know about the eigenvalues of A and A^{-1}?
10. **Hint.** Look at some examples of 2×2 diagonal matrices.

CHAPTER TEST B

1. **(c) Hint.** $A^7 = XD^7X^{-1}$
2. **Hint.** $\mathrm{tr}(A) = 4$
4. **Hint.** Show that $\dim N(A - aI) < 3$.
5. **(a) Hint.** Show that A can be reduced to upper triangular form U using only row operation III and that the diagonal entries of U are all positive.
   (b) Hint. The elimination process in part (a) should yield an LU factorization of A. The upper triangular matrix U should factor into a product DL^T.
7. **Hint.** A is diagonalizable, so e^{tA} can be computed as a product $Xe^{tD}X^{-1}$.
8. **(c) Solution.** Set
$$\mathbf{u}_1 = \frac{1}{\|\mathbf{x}_1\|}$$
and then use the Gram-Schmidt process to transform the eigenspace basis $\{\mathbf{x}_2, \mathbf{x}_3, \mathbf{x}_4\}$ into an orthonormal basis $\{\mathbf{u}_2, \mathbf{u}_3, \mathbf{u}_4\}$.
9. **(b) Hint.** If $\mathbf{z} = c_1\mathbf{u}_1 + c_2\mathbf{u}_2$, then
$$\|\mathbf{z}\|_2^2 = |c_1|^2 + |c_2|^2$$
10. **Hint.** Show that B and C are both symmetric. What can you conclude about the eigenvalues of these matrices?

Chapter 7

Numerical Linear Algebra

1 | FLOATING-POINT NUMBERS

Overview

In this chapter we consider computer methods for solving linear algebra problems. To understand these methods, one should be familiar with the type of number system used by the computer. When data are read into the computer, they are translated into the finite number system of the computer. This translation will usually involve some roundoff error. Further roundoff errors will occur when the algebraic operations of the algorithm are carried out. In this section we learn about the floating point number system used by computers and how roundoff error effects algorithms.

121

Important Concepts

1. **Numerical Stability.** An algorithm is said to be *stable* if it will produce a good approximation to a slightly perturbed version of the problem.

2. **Floating point numbers.** A *floating-point number* in base b is a number of the form
$$\pm\left(\frac{d_1}{b} + \frac{d_2}{b^2} + \cdots + \frac{d_t}{b^t}\right) \times b^e$$
where $t, d_1, d_2, \ldots, d_t, b, e$ are all integers and
$$0 \le d_i \le b-1 \qquad i = 1, \ldots, t$$
The integer t refers to the number of digits and this depends on the word length of the computer. The exponent e is restricted to be within certain bounds, $L \le e \le U$, which also depend on the particular computer. The floating point representation of a real number x is denoted $fl(x)$.

3. **Roundoff error.** If x is a real number and x' is an approximation to x, then the difference $x' - x$ is called the *absolute error* and the quotient $(x' - x)/x$ is called the *relative error*.

4. **Machine epsilon.** The *machine epsilon* is the smallest floating-point number ϵ for which
$$fl(1 + \epsilon) > 1$$

Important Results

▶ If two floating point numbers a and b are close, then there will be a loss of significant digits when there difference is computed in finite precision arithmetic. In such a case the relative error in approximating $a - b$ by $fl(a - b)$ will be much greater than the relative errors in either a or b.

Exercises: Solutions and Hints

2. (c) **Solution.** The three digit decimal floating point representation of 0.01277 if given by
$$fl(0.01277) = 0.128 \times 10^{-1}$$
The absolute error is
$$fl(0.01277) - 0.01277 = 0.128 \times 10^{-1} - 0.01277 = 0.0003$$
The relative error is
$$\delta = \frac{fl(0.01277) - 0.01277}{0.01277} = \frac{0.0003}{0.01277} \approx 0.0235$$

3. (c) **Solution.** To write the number in terms of powers of 2 we proceed as follows
$$\begin{aligned} 9.872 &= 2^3 + 1.872 \\ &= 2^3 + 0 \cdot 2^1 + 1.872 \\ &= 2^3 + 0 \cdot 2^1 + 2^0 + 0.872 \end{aligned}$$

$$= 2^3 + 0 \cdot 2^1 + 2^0 + 2^{-1} + 0.372$$
$$= 2^3 + 0 \cdot 2^1 + 2^0 + 2^{-1} + 2^{-2} + 0.122$$

We have now expressed 9.872 as a sum of 5 consecutive powers of 2 plus a remainder. Since the remainder 0.122 is less than $2^{-3} = 0.125$ we round off to 0. We then have

$$9.872 \approx 1 \cdot 2^3 + 0 \cdot 1 \cdot 2^1 + 1 \cdot 2^0 + 1 \cdot 2^{-1} + 1 \cdot 2^{-2}$$

The 5 digit base 2 floating point representation of 9.872 is then

$$fl(9.872) = (0.10111)_2 \times 2^3$$

4. (c) **Solution.** First the two numbers, call them a and b, must be represented as four-digit decimal floating point numbers.

$$fl(a) = fl(0.12347) = 0.1235 \times 10^0$$
$$fl(b) = fl(0.12342) = 0.1234 \times 10^0$$

The exact difference d and the floating point difference d' are given by

$$d = a - b = 0.00005 \qquad \text{and} \qquad d' = fl(fl(a) - fl(b)) = 0.1000 \times 10^{-4}$$

The absolute and relative errors are

$$e = d' - d = 0.00005 \qquad \text{and} \qquad \delta = \frac{d' - d}{d} = 1.00$$

There is a 100% error in the calculated difference.

---

2 | GAUSSIAN ELIMINATION

Overview

The most computationally efficient direct method for solving linear systems is Gaussian elimination. If the coefficient matrix A for a linear system can be reduced to triangular form using only row operation III, then A can be factored into a product LU. The system $A\mathbf{x} = \mathbf{b}$ can then be solved using forward and back substitution.

Important Concepts

1. **Gaussian elimination without interchanges.** The process of reducing a matrix to upper triangular form using only row operation III is known as *Gaussian elimination without interchanges.*

2. **Forward and back substitution.** If A has triangular factorization LU, then a system $A\mathbf{x} = \mathbf{b}$ can be solved by first using *forward substitution* to solve a lower triangular system $L\mathbf{y} = \mathbf{b}$ and then using *back substitution* to $A\mathbf{x} = \mathbf{y}$.

Important Results

▶ ALGORITHM 7.2.1 (Gaussian Elimination without Interchanges). *This algorithm is given in the textbook.*

▶ ALGORITHM 7.2.2 (Forward and Back Substitution). *This algorithm is given in the textbook.*

Exercises: Solutions and Hints

4. (a) **Solution.**
    (i) To compute $(A(\mathbf{xy}^T))B$ we must first we compute the product $\mathbf{xy}^T$. If $C = \mathbf{xy}^T$, then its (i, j) entry is $c_{ij} = x_i y_j$. To compute this entry we must do 1 multiplication an 0 additions. Since C has n^2 entries, it can be computed using n^2 multiplications and 0 additions. Next we compute $D = AC$. The (i, j) entry of D is

    $$d_{ij} = a_{i1}c_{1j} + a_{i2}c_{2j} + \cdots + a_{in}c_{nj}$$

    This computation requires n multiplications and $n - 1$ additions. Since D has mn entries, the computation of the product $D = AC$ requires mn^2 multiplications and $mn(n-1)$ additions. Finally we must compute $E = DB$. The (i, j) entry of E is

    $$e_{ij} = d_{i1}b_{1j} + d_{i2}b_{2j} + \cdots + d_{in}b_{nj}$$

    This computation requires n multiplications and $n-1$ additions. Since E has mr entries, the computation of the product $E = DB$ requires mrn multiplications and $mr(n-1)$ additions. The total number of operations to compute $(A(\mathbf{xy}^T))B$ is

    Multiplications: $n^2 + mn^2 + mrn = n(n + mn + mr)$
    Additions: $0 + mn(n - 1) + mr(n - 1) = m(n + r)(n - 1)$

    (ii) To compute $(A\mathbf{x})(\mathbf{y}^T B)$ we must first we compute $\mathbf{c} = A\mathbf{x}$. The ith entry of $\mathbf{c}$ is

    $$c_i = a_{i1}x_1 + a_{i2}x_2 + \cdots + a_{in}\mathbf{x}_n$$

    The computation of c_i requires n multiplications and $n - 1$ additions. Since $\mathbf{c}$ has m entries, it follows that the computation of $\mathbf{c}$ will require mn multiplications and $m(n-1)$ additions. Next we compute $\mathbf{d}^T = \mathbf{y}^T B$. The ith entry of $\mathbf{d}^T$ is

    $$d_i = y_1 b_{1i} + y_2 b_{2i} + \cdots + y_n b_{ni}$$

    The computation of d_i requires n multiplications and $n - 1$ additions. Since $\mathbf{d}^T$ has r entries, it follows that the computation of $\mathbf{d}$ will require nr multiplications and $(n - 1)r$ additions. Finally to compute $(A\mathbf{x})(\mathbf{y}^T B)$, we must multiply $E = \mathbf{cd}^T$. The (i, j) entry of this matrix is $e_{ij} = c_i d_j$. The computation of each entry of this product requires only 1 multiplication and 0 additions. Since E has mr entries the computation of E will require mr multiplications and 0 additions. The total number of operations to compute $(A\mathbf{x})(\mathbf{y}^T B)$ is

    Multiplications: $mn + nr + mr$
    Additions: $m(n - 1) + (n - 1)r + 0 = (n - 1)(m + r)$

7. Solution. Algorithm for solving $LDL^T\mathbf{x} = \mathbf{b}$

For $k = 1, \ldots, n$

$$\text{Set } y_k = b_k - \sum_{i=1}^{k-1} \ell_{ki}y_i$$

End (For Loop)

For $k = 1, \ldots, n$

$$\text{Set } z_k = y_k/d_{ii}$$

End (For Loop)

For $k = n - 1, \ldots, 1$

$$\text{Set } x_k = z_k - \sum_{j=k+1}^{n} \ell_{jk}x_j$$

End (For Loop)

10. Hint. The operation count for the multiplication $A^{-1}\mathbf{b}$ is the same as the operation count for Exercise 3(a). In determining the operation counts for doing forward and back substitution the following formula is useful.

$$1 + 2 + \cdots + n = \frac{n(n+1)}{2}$$

11. Solution. If

$$A(E_1 E_2 E_3) = L$$

then

$$A = L(E_1 E_2 E_3)^{-1} = LU$$

The elementary matrices are chosen to zero out the $(1,2)$, $(1,3)$ and $(2,3)$ entries of A. The inverse elementary matrices E_1^{-1}, E_2^{-1}, E_3^{-1} will each be upper triangular with ones on the diagonal. Indeed,

$$E_1^{-1} = \begin{pmatrix} 1 & \frac{a_{12}}{a_{11}} & 0 \\ 0 & 1 & 0 \\ 0 & 0 & 1 \end{pmatrix} \quad E_2^{-1} = \begin{pmatrix} 1 & 0 & \frac{a_{13}}{a_{11}} \\ 0 & 1 & 0 \\ 0 & 0 & 1 \end{pmatrix} \quad E_3^{-1} = \begin{pmatrix} 1 & 0 & 0 \\ 0 & 1 & \frac{a_{23}}{a_{22}^{(1)}} \\ 0 & 0 & 1 \end{pmatrix}$$

where $a_{22}^{(1)} = a_{22} - \frac{a_{12}}{a_{11}}$. If we let

$$u_{12} = \frac{a_{12}}{a_{11}}, \quad u_{13} = \frac{a_{13}}{a_{11}}, \quad u_{23} = \frac{a_{23}}{a_{22}^{(1)}}$$

then

$$U = E_3^{-1}E_2^{-1}E_1^{-1} = \begin{pmatrix} 1 & u_{12} & u_{13} \\ 0 & 1 & u_{23} \\ 0 & 0 & 1 \end{pmatrix}$$

3 | PIVOTING STRATEGIES

Overview

In this section we present an algorithm for Gaussian elimination with row interchanges. At each step of the algorithm it will be necessary to choose a pivotal row. One can avoid unnecessarily large accumulations of roundoff error by using a strategy called partial pivoting for choosing the pivotal rows. If P is the permutation matrix corresponding to the row interchanges in the elimination process, then the matrix PA has an LU factorization.

Important Concepts

1. **Partial pivoting.** At the ith step of the reduction process, there are $n-i+1$ candidates for the pivot element:

$$a_{p(i),i}, a_{p(i+1),i}, \ldots, a_{p(n),i}$$

where $\mathbf{p}$ is a vector that keeps track of the ordering of the rows. The *partial pivoting* strategy chooses the candidate $a_{p(j),i}$ with maximum modulus

$$|a_{p(j),i}| = \max_{i \leq k \leq n} |a_{p(k),i}|$$

The ith and jth entries of $\mathbf{p}$ are then interchanged. The pivot element $a_{p(i),i}$ has the property

$$|a_{p(i),i}| \geq |a_{p(k),i}|$$

for $k = i+1, \ldots, n$. Thus the multipliers will all satisfy

$$|m_{p(k),i}| = \left| \frac{a_{p(k),i}}{a_{p(i),i}} \right| \leq 1$$

and this prevents any significant build up of roundoff error.

2. **Complete pivoting.** The *complete pivoting* strategy works the same way as partial pivoting except that the pivot element is chosen to be the element of maximum absolute value among all the elements in the remaining rows and columns. After the pivot element is chosen it is necessary to interchange rows and also to interchange columns.

Important Results

▶ If A is reduced to triangular form using only row operations I and III and P is the permutation matrix determined by the row interchanges in the elimination process, then PA has a triangular factorization LU.

▶ ALGORITHM 7.3.1 (Gaussian elimination with interchanges). *See the textbook for the details of this algorithm.*

Exercises: Solutions and Hints

6. Solution.

(a) The entry of A with the maximum absolute value is $a_{32} = 8$. This is our choice for the first pivot element. We use it to eliminate the other two entries in its column.

$$\left(\begin{array}{ccc|c} 5 & 4 & 7 & 2 \\ 2 & -4 & 3 & -5 \\ 2 & 8 & 6 & 4 \end{array}\right) \rightarrow \left(\begin{array}{ccc|c} 4 & 0 & 4 & 0 \\ 3 & 0 & 6 & -3 \\ 2 & 8 & 6 & 4 \end{array}\right)$$

Next, if we exclude the third row and second column the largest entry of the resulting matrix is the 6 in the $(2,3)$ position. We use this as a pivot element to eliminate the entry in the $(1,3)$ position.

$$\left(\begin{array}{ccc|c} 4 & 0 & 4 & 0 \\ 3 & 0 & 6 & -3 \\ 2 & 8 & 6 & 4 \end{array}\right) \rightarrow \left(\begin{array}{ccc|c} 2 & 0 & 0 & 2 \\ 3 & 0 & 6 & -3 \\ 2 & 8 & 6 & 4 \end{array}\right)$$

We can now successively solve for x_1, x_3, and x_2.

$$2x_1 = 2 \qquad x_1 = 1$$
$$3 + 6x_3 = -3 \qquad x_3 = -1$$
$$2 + 8x_2 - 6 = 4 \qquad x_2 = 1$$

$$\mathbf{x} = (1,\ 1,\ -1)^T$$

(b) The pivot rows were $3, 2, 1$ and the pivot columns were $2, 3, 1$. Therefore

$$P = \begin{pmatrix} 0 & 0 & 1 \\ 0 & 1 & 0 \\ 1 & 0 & 0 \end{pmatrix} \qquad \text{and} \qquad Q = \begin{pmatrix} 0 & 0 & 1 \\ 1 & 0 & 0 \\ 0 & 1 & 0 \end{pmatrix}$$

Rearranging the rows and columns of the reduced matrix from part (a), we get

$$U = \begin{pmatrix} 8 & 6 & 2 \\ 0 & 6 & 3 \\ 0 & 0 & 2 \end{pmatrix}$$

The matrix L is formed using the multipliers $-\frac{1}{2}, \frac{1}{2}, \frac{2}{3}$.

$$L = \begin{pmatrix} 1 & 0 & 0 \\ -\frac{1}{2} & 1 & 0 \\ \frac{1}{2} & \frac{2}{3} & 1 \end{pmatrix}$$

(c) To solve the system $A\mathbf{x} = \mathbf{c}$ we set $\mathbf{d} = P\mathbf{c}$ and multiply both sides of the matrix equation by P.

$$PA\mathbf{x} = P\mathbf{c} = \mathbf{d} = \begin{pmatrix} 2 \\ -4 \\ 5 \end{pmatrix}$$

If we now set $\mathbf{y} = Q^T\mathbf{x}$, then $\mathbf{x} = Q\mathbf{y}$ and it follows that

$$PA\mathbf{x} = PAQ\mathbf{y}$$

Since $PAQ = LU$ we have transformed the system to

$$LU\mathbf{y} = \mathbf{d}$$

We can now use forward and back substitution to solve for $\mathbf{y}$.

Forward Substitution

Solve $L\mathbf{y} = \mathbf{d}$

$$\left(\begin{array}{ccc|c} 1 & 0 & 0 & 2 \\ -\frac{1}{2} & 1 & 0 & -4 \\ \frac{1}{2} & \frac{2}{3} & 1 & 5 \end{array} \right) \qquad \begin{array}{l} y_1 = 2 \\ y_2 = -3 \\ y_3 = 6 \end{array}$$

Back substitution

Solve $U\mathbf{z} = \mathbf{y}$

$$\left(\begin{array}{ccc|c} 8 & 6 & 2 & 2 \\ 0 & 6 & 3 & -3 \\ 0 & 0 & 2 & 6 \end{array} \right) \qquad \begin{array}{l} z_1 = 1 \\ z_2 = -2 \\ z_3 = 3 \end{array}$$

Finally, to obtain the solution to the system we set $\mathbf{x} = Q\mathbf{z}$

$$\mathbf{x} = \begin{pmatrix} 0 & 0 & 1 \\ 1 & 0 & 0 \\ 0 & 1 & 0 \end{pmatrix} \begin{pmatrix} 1 \\ -2 \\ 3 \end{pmatrix} = \begin{pmatrix} 3 \\ 1 \\ -2 \end{pmatrix}$$

4 | MATRIX NORMS AND CONDITION NUMBERS

Overview

In this section we are concerned with the accuracy of computed solutions to linear systems. How accurate can we expect the computed solutions to be, and how can one test their accuracy? The answer to these questions depends largely on how sensitive the coefficient matrix of the system is to small changes. The sensitivity of the matrix can be measured in terms of its *condition number*. The condition number of a nonsingular matrix is defined in terms of its norm and the norm of its inverse. In this section we study the most important families of matrix norms. We then make use of matrix norms and condition numbers to derive error bounds for computed solutions to linear systems.

Important Concepts

1. **Matrix norm.** A *matrix norm* is a norm defined on the vector space $R^{m \times n}$. It must satisfy the three conditions that define norms in general.
   (i) $\|A\| \geq 0$ and $\|A\| = 0$ if and only if $A = O$
   (ii) $\|\alpha A\| = |\alpha| \|A\|$
   (iii) $\|A + B\| \leq \|A\| + \|B\|$
   The families of norms that turn out to be most useful also satisfy the additional property
   (iv) $\|AB\| \leq \|A\| \|B\|$

2. **Frobenius norm.** The *Frobenius norm* of a matrix A is computed by taking the square root of the sum of the squares of all of its entries.

$$\|A\|_F = \left(\sum_{j=1}^{n} \sum_{i=1}^{m} a_{ij}^2 \right)^{1/2}$$

3. **Compatible norms.** A matrix norm $\| \cdot \|_M$ on $R^{m \times n}$ and a vector norm $\| \cdot \|_V$ on R^n are said to be *compatible* if

$$\|A\mathbf{x}\|_V \leq \|A\|_M \|\mathbf{x}\|_V$$

for every $\mathbf{x} \in R^n$.

4. **Subordinate matrix norms.** For each family of vector norms $\| \cdot \|$, one can then define a family of matrix norms by

$$\|A\| = \max_{\mathbf{x} \neq \mathbf{0}} \frac{\|A\mathbf{x}\|}{\|\mathbf{x}\|}$$

If the matrix norm is defined in this way, then we say it is *subordinate* to the vector norm $\| \cdot \|$.

5. **Ill-conditioned and well-conditioned matrices.** A matrix A is said to be *ill-conditioned* if relatively small changes in the entries of A can cause relatively large changes in the solutions to $A\mathbf{x} = \mathbf{b}$. A is said to be *well-conditioned* if relatively small changes in the entries of A result in relatively small changes in the solutions to $A\mathbf{x} = \mathbf{b}$.

6. **Condition numbers.** The number $\|A\| \, \|A^{-1}\|$ is called the *condition number* of A and is denoted by $\text{cond}(A)$.

Important Theorems and Results

▶ THEOREM 7.4.1. *If the family of matrix norms $\| \cdot \|_M$ is subordinate to the family of vector norms $\| \cdot \|_V$, then $\| \cdot \|_M$ and $\| \cdot \|_V$ are compatible and the matrix norms $\| \cdot \|_M$ satisfy property* (iv).

▶ THEOREM 7.4.2. *If A is an $m \times n$ matrix, then*

$$\|A\|_1 = \max_{1 \leq j \leq n} \left(\sum_{i=1}^{m} |a_{ij}| \right)$$

and

$$\|A\|_\infty = \max_{1 \leq i \leq m} \left(\sum_{j=1}^{n} |a_{ij}| \right)$$

▶ THEOREM 7.4.3. *If A is an $m \times n$ matrix with singular value decomposition $U\Sigma V^T$, then*

$$\|A\|_2 = \sigma_1 \qquad \text{(the largest singular value)}$$

▶ COROLLARY 7.4.4. *If $A = U\Sigma V^T$ is nonsingular, then*

$$\|A^{-1}\|_2 = \frac{1}{\sigma_n}$$

▶ Error Bounds. If $\mathbf{x}$ is the exact solution to $A\mathbf{x} = \mathbf{b}$ and $\mathbf{x}'$ is the computed solution, then

$$\frac{1}{\text{cond}(A)} \frac{\|\mathbf{r}\|}{\|\mathbf{b}\|} \leq \frac{\|\mathbf{x}' - \mathbf{x}\|}{\|\mathbf{x}\|} \leq \text{cond}(A) \frac{\|\mathbf{r}\|}{\|\mathbf{b}\|}$$

where $\mathbf{r} = \mathbf{b} - A\mathbf{x}'$.

▶ **Perturbation Error Bound.** If $\mathbf{x}$ is the exact solution to $A\mathbf{x} = \mathbf{b}$ and $\mathbf{x}'$ is the exact solution to the perturbed system $(A + E)\mathbf{x} = \mathbf{b}$, then

$$\frac{\|\mathbf{x} - \mathbf{x}'\|}{\|\mathbf{x}'\|} \leq \text{cond}(A)\frac{\|E\|}{\|A\|}$$

Exercises: Solutions and Hints

3. Hint. The value of $\|A\|_2$ is 1 since A has singular values $\sigma_1 = 1$ and $\sigma_2 = 0$. To show that the $\|A\|_2 = 1$ using the definition of the 2-norm you must show that

$$\max_{\mathbf{x} \neq \mathbf{0}} \frac{\|A\mathbf{x}\|}{\|\mathbf{x}\|} = \max_{\mathbf{x} \neq \mathbf{0}} \frac{\sqrt{x_1^2}}{\sqrt{x_1^2 + x_2^2}} = 1$$

To show this consider the 2 cases: (i) $x_2 \neq 0$ and (ii) $x_2 = 0$.

5. Hint. If

$$|d_{kk}| = \max_{1 \leq i \leq n}(|d_{ii}|)$$

you can establish the result by either using the definition and showing

$$\max_{\mathbf{x} \neq \mathbf{0}} \frac{\|D\mathbf{x}\|}{\|\mathbf{x}\|} = |d_{kk}|$$

or you can make use of Theorem 7.4.3 and show that the largest singular value of D is $|d_{kk}|$.

11. (b) Hint. Construct the vector $\mathbf{x}$ in the same way you did for part (b) of Exercise 10.

14. Hint. $\|A\|_F = \|\Sigma\|_F$.

17. Hint. If $\mathbf{x}$ is a unit eigenvector belonging to the eigenvalue λ, then $|\lambda| = \|\lambda\mathbf{x}\|$.

19. Parts (a) and (b). Hint. For any vector $\mathbf{y} \in R^n$

$$\|\mathbf{y}\|_\infty \leq \|\mathbf{y}\|_2 \leq n^{1/2}\|\mathbf{y}\|_\infty$$

(c) **Hint.** Use the results from parts (a) and (b).

20. Hint. Let A be a symmetric matrix with orthonormal eigenvectors $\mathbf{u}_1, \ldots, \mathbf{u}_n$. If $\mathbf{x} \in R^n$ then by Theorem 5.5.2

$$\mathbf{x} = c_1\mathbf{u}_1 + c_2\mathbf{u}_2 + \cdots + c_n\mathbf{u}_n$$

where $c_i = \mathbf{u}_i^T\mathbf{x}$, $i = 1, \ldots, n$. Make use of Parseval's formula in parts (a) and (b). For part (c) find a vector $\mathbf{x}$ such that

$$\frac{\|A\mathbf{x}\|_2}{\|\mathbf{x}\|_2} = \max_{1 \leq i \leq n} |\lambda_i|$$

24. Hint. Make use of the result from Exercise 8(a).

26. Hint. Express $\|A\|_2$, $\text{cond}_2(A)$, and $\|A\|_F$ in terms of the singular values of A.

33. (a) Hint. This can be shown using the definition of the 2-norm.
   (b) **Hint.** The matrix $Q^{-1} = Q^T$ is also orthogonal.

34. (a) **Hint.** Show first that

$$\|QA\mathbf{x}\|_2 = \|A\mathbf{x}\|_2$$

   (b) **Hint.** For each nonzero vector $\mathbf{x}$ in R^n, set $\mathbf{y} = V\mathbf{x}$. Since V is nonsingular it follows that $\mathbf{y}$ is nonzero. Conversely, if we are given a nonzero vector $\mathbf{y}$, then $\mathbf{x} = V^T\mathbf{y}$ is a nonzero vector and $\mathbf{y} = V\mathbf{x}$.

35. (b) **Hint.** You need to find vectors $\mathbf{x}_1$ and $\mathbf{y}_1$ for which equality holds in part (a).

36. **Hint.** If $\mathbf{y} = V^T\mathbf{x}$, show that

$$\|A\mathbf{x}\|_2 = \|\Sigma\mathbf{y}\|_2 \geq \sigma_n\|\mathbf{x}\|_2$$

   Show also that if we take $\mathbf{x} = \mathbf{v}_n$ then

$$\|A\mathbf{x}\|_2 = \sigma_n = \sigma_n\|\mathbf{x}\|_2$$

---

5 ORTHOGONAL TRANSFORMATIONS

Overview

Orthogonal transformations are one of the most important tools in numerical linear algebra. In this section we study special classes of orthogonal transformations that can be used to zero out entries of a matrix.

Important Concepts

1. **Elementary orthogonal matrix.** An *elementary orthogonal matrix*, is a matrix of the form

$$Q = I - 2\mathbf{u}\mathbf{u}^T$$

   where $\mathbf{u} \in R^n$ and $\|\mathbf{u}\|_2 = 1$.

2. **Householder transformation.** A *Householder transformation* is an elementary orthogonal matrix that has the effect of zeroing out the last $n - 1$ entries of a vector $\mathbf{x} \in R^n$.

3. **Plane rotation.** A *plane rotation* acting on R^2 is a matrix of the form

$$R = \begin{pmatrix} \cos\theta & -\sin\theta \\ \sin\theta & \cos\theta \end{pmatrix}$$

A *plane rotation* for R^n is a matrix of the form

$$
R = \begin{pmatrix}
1 & 0 & \cdots & 0 & \cdots & 0 & \cdots & 0 \\
0 & 1 & \cdots & 0 & \cdots & 0 & \cdots & 0 \\
\vdots & & & & & & & \\
0 & 0 & \cdots & \cos\theta & \cdots & -\sin\theta & \cdots & 0 \\
\vdots & & & & & & & \\
0 & 0 & \cdots & \sin\theta & \cdots & \cos\theta & \cdots & 0 \\
\vdots & & & & & & & \\
0 & 0 & \cdots & 0 & \cdots & 0 & \cdots & 1
\end{pmatrix}
$$

4. Givens transformation. A *Givens transformation* or *Givens reflection* acting on R^2 is a matrix of the form

$$
G = \begin{pmatrix}
\cos\theta & \sin\theta \\
\sin\theta & -\cos\theta
\end{pmatrix}
$$

A *Givens transformation* for R^n is a matrix of the form

$$
G = \begin{pmatrix}
1 & 0 & \cdots & 0 & \cdots & 0 & \cdots & 0 \\
0 & 1 & \cdots & 0 & \cdots & 0 & \cdots & 0 \\
\vdots & & & & & & & \\
0 & 0 & \cdots & \cos\theta & \cdots & \sin\theta & \cdots & 0 \\
\vdots & & & & & & & \\
0 & 0 & \cdots & \sin\theta & \cdots & -\cos\theta & \cdots & 0 \\
\vdots & & & & & & & \\
0 & 0 & \cdots & 0 & \cdots & 0 & \cdots & 1
\end{pmatrix}
$$

Important Results

▶ Householder elimination. To construct a Householder matrix that zeros out the last $n - 1$ entries of a vector $\mathbf{x} \in R^n$ set

$$\alpha = \|\mathbf{x}\|_2, \qquad \beta = \alpha(\alpha - x_1)$$
$$\mathbf{v} = (x_1 - \alpha, x_2, \ldots, x_n)^T$$
$$H = \tfrac{1}{\beta}\mathbf{v}\mathbf{v}^T$$

The matrix H has the property that

$$H\mathbf{x} = \alpha\mathbf{e}_1$$

where $\mathbf{e}_1$ is the first column of the $n \times n$ identity matrix.

▶ Elimination using plane rotations. To construct a rotation matrix R that zeros out the second entry of a vector $\mathbf{x}$ in R^2 set

$$r = \sqrt{x_1^2 + x_2^2}, \quad c = \frac{x_1}{r}, \quad s = \frac{x_2}{r}$$

and

$$R = \begin{pmatrix} c & -s \\ s & c \end{pmatrix}$$

The matrix R has the property that

$$R\mathbf{x} = \begin{pmatrix} r \\ 0 \end{pmatrix}$$

To construct a rotation matrix R that acts only on the ith and jth entries of a vector $\mathbf{x}$ in R^n and zeros out the jth entry set

$$r = \sqrt{x_i^2 + x_j^2}, \quad c = \frac{x_i}{r}, \quad s - \frac{x_j}{r}$$
$$r_{ii} = r_{jj} = c, \quad r_{ij} = -s; \quad r_{ji} = s$$

The remaining entries of R are the same as those of the $n \times n$ identity matrix.

▶ Elimination using Givens transformations. To construct a Givens transformation G that zeros out the second entry of a vector $\mathbf{x}$ in R^2 set

$$r = \sqrt{x_1^2 + x_2^2}, \quad c = \frac{x_1}{r}, \quad s = \frac{x_2}{r}$$

and

$$G = \begin{pmatrix} c & s \\ s & -c \end{pmatrix}$$

The matrix G has the property that

$$G\mathbf{x} = \begin{pmatrix} r \\ 0 \end{pmatrix}$$

To construct a Givens transformation G that acts only on the ith and jth entries of a vector $\mathbf{x}$ in R^n and zeros out the jth entry set

$$r = \sqrt{x_i^2 + x_j^2}, \quad c = \frac{x_i}{r}, \quad s = \frac{x_j}{r}$$

$$g_{ii} = c, \quad g_{ij} = g_{ji} = s, \quad g_{jj} = -c$$

The remaining entries of G are the same as those of the $n \times n$ identity matrix.

▶ **Householder QR factorization.** If A is an $m \times n$ matrix of rank n, then you can successively apply Householder transformations to A and obtain a factorization

$$A = QR$$

where Q is an $m \times m$ orthogonal matrix that is a product of $n - 1$ Householder transformations (i.e., $Q = H_1 H_2 \cdots H_{n-1}$) and R is an $m \times n$ matrix that is upper triangular with nonzero diagonal entries.

Exercises: Solutions and Hints

7. (b) **Solution** The first column of the coefficient matrix is

$$\mathbf{a}_1 = \begin{pmatrix} 1 \\ 1 \end{pmatrix}$$

and $\|\mathbf{a}_1\| = \sqrt{2}$. We need to find a Givens transformation G such that

$$G\mathbf{a}_1 = \begin{pmatrix} \sqrt{2} \\ 0 \end{pmatrix}$$

Set

$$r = \sqrt{a_{11}^2 + a_{21}^2} = \sqrt{2}, \quad c = \frac{a_{11}}{r} = \frac{1}{\sqrt{2}}, \quad s = \frac{a_{12}}{r} = \frac{1}{\sqrt{2}}$$

and set

$$G = \begin{pmatrix} c & s \\ s & -c \end{pmatrix} = \begin{pmatrix} \frac{1}{\sqrt{2}} & \frac{1}{\sqrt{2}} \\ \frac{1}{\sqrt{2}} & -\frac{1}{\sqrt{2}} \end{pmatrix}$$

To obtain an upper triangular system we multiply the augmented matrix $(A|\mathbf{b})$ by G.

$$G(A|\mathbf{b}) = (GA \mid G\mathbf{b}) = \begin{pmatrix} \sqrt{2} & 3\sqrt{2} & 3\sqrt{2} \\ 0 & \sqrt{2} & 2\sqrt{2} \end{pmatrix}$$

The solution $\mathbf{x}$ is obtained using back substitution.

$$\mathbf{x} = \begin{pmatrix} -3 \\ 2 \end{pmatrix}$$

12. (b) **Hint.** $\mathbf{u}^T\mathbf{x}$ is a 1×1 matrix, so the matrix product $2\mathbf{u}\mathbf{u}^T\mathbf{x}$ can be expressed as a scalar multiple of $\mathbf{x}$.

$$2\mathbf{u}\mathbf{u}^T\mathbf{x} = (2\mathbf{u}^T\mathbf{x})\mathbf{u}$$

Use the result from part (a) to show that

$$2\mathbf{u}^T\mathbf{x} = \|\mathbf{x} - \mathbf{y}\|$$

15. (a) **Hint.** Let $Q = Q_1^T Q_2 = R_1 R_2^{-1}$. Show that Q must be both orthogonal and upper triangular. If a matrix has both of these properties, what can you conclude?

16. Hint. Let H_1 and H_2 be Householder matrices with the properties

$$H_1\mathbf{x} = \|\mathbf{x}\|\, \mathbf{e}_1^{(m)} \qquad \text{and} \qquad H_2\mathbf{y} = \|\mathbf{y}\|\, \mathbf{e}_1^{(n)}$$

where $\mathbf{e}_1^{(m)}$ is the first column vector of the $m \times m$ identity matrix and $\mathbf{e}_1^{(n)}$ is the first column vector of the $n \times n$ identity matrix.

17. Hint. Read the Important Result that is listed for Section 1 of this chapter.

6 THE EIGENVALUE PROBLEM

Overview

In this section we are concerned with numerical methods for computing the eigenvalues and eigenvectors of an $n \times n$ matrix A. The first method we study is called the power method. It involves successively applying powers of A to an initial vector and scaling the product at each step. If no two eigenvalues have the same absolute value, then the power method can be used to compute the eigenvalues of a matrix one at a time. A second and more sophisticated method, the QR algorithm, computes all of the eigenvalues and eigenvectors at the same time.

Important Concepts

1. **Dominant eigenvalue.** By the *dominant eigenvalue* we mean an eigenvalue λ_1 satisfying $|\lambda_1| > |\lambda_i|$ for $i = 2, \ldots, n$.
2. **Power Method.** The *power method* is an iterative method for finding the dominant eigenvalue of a matrix and a corresponding eigenvector.
3. **Deflation.** Once the dominant eigenvalue of an $n \times n$ matrix has has been found, the remaining $n - 1$ eigenvalues of A are the eigenvalues of an $(n-1) \times (n-1)$ matrix A_1. The matrix A_1 is determined by a process called *deflation* which involves applying a similarity transformation to A using a single Householder matrix.

4. **Upper Hessenberg form.** A matrix A is said to be in *upper Hessenberg form* if $a_{ij} = 0$ whenever $i \geq j + 2$.

5. **QR algorithm.** The QR *algorithm* is an iterative method involving orthogonal similarity transformations. It will converge whether or not A has a dominant eigenvalue and it calculates all of the eigenvalues at the same time.

Important Theorems

▶ THEOREM 7.6.1. *Let A be an $n \times n$ matrix with n linearly independent eigenvectors and let X be a matrix that diagonalizes A.*

$$X^{-1}AX = D = \begin{pmatrix} \lambda_1 & & & \\ & \lambda_2 & & \\ & & \ddots & \\ & & & \lambda_n \end{pmatrix}$$

If $A' = A + E$ and λ' is an eigenvalue of A', then

$$\min_{1 \leq i \leq n} |\lambda' - \lambda_i| \leq \text{cond}_2(X)\|E\|_2$$

▶ **Power Method.** If an $n \times n$ matrix A has a dominant eigenvalue λ_1, then the power method is an iterative process to compute an eigenvector $\mathbf{y}$ belonging to λ_1. To start, $\mathbf{u}_0$ can be any nonzero vector in R^n. Two sequences $\{\mathbf{v}_k\}$ and $\{\mathbf{u}_k\}$ are then defined recursively. Once $\mathbf{u}_k$ has been determined, the vectors $\mathbf{v}_{k+1}$ and $\mathbf{u}_{k+1}$ are calculated as follows:

1. Set $\mathbf{v}_{k+1} = A\mathbf{u}_k$.
2. Find the coordinate j_{k+1} of $\mathbf{v}_{k+1}$ of maximum modulus.
3. Set $\mathbf{u}_{k+1} = (1/v_{j_{k+1}})\mathbf{v}_{k+1}$.

If the initial vector $\mathbf{u}_0$ is not in $N(A - \lambda_1 I)^\perp$, then $\mathbf{u}_k \to \mathbf{y}$ and $\mathbf{v}_k \to \lambda_1\mathbf{y}$.

▶ QR **algorithmn.** The algorithm is outlined in the textbook. The basic idea is to first use a similarity transformation to reduce the $n \times n$ matrix A to upper Hessenberg form U. This can be done using $n - 2$ Householder matrices

$$H_{n-2} \cdots H_2 H_1 A H_1 H_2 \cdots H_{n-2} = U$$

If $Q = H_1 H_2 \cdots H_{n-2}$, then Q is orthogonal and $Q^T AQ = U$. The second part of the algorithm applies a sequence of Givens transformations to each side of U until the result converges to a block upper triangular matrix T where the diagonal blocks of T are all 1×1 or 2×2. Thus for k sufficiently large

$$G_k \cdots G_2 G_1 U G_1 G_2 \cdots G_k \approx T$$

Note that if

$$W = QG_1G_2 \cdots G_k$$

then W is orthogonal and

$$W^{-1}AW = W^T AW = T$$

So A and T are similar and hence have the same eigenvalues. The eigenvalues of T are just the eigenvalues of its diagonal blocks. In the case that A is symmetric, the matrix U will be symmetric tridiagonal and the matrix T will be diagonal.

Exercises: Solutions and Hints

3. (a) **Solution.** $\mathbf{v}_1 = A\mathbf{u}_0 = \begin{pmatrix} 3 \\ -2 \end{pmatrix}$ $\mathbf{u}_1 = \frac{1}{3}\mathbf{v}_1 = \begin{pmatrix} 1 \\ -2/3 \end{pmatrix}$

$\mathbf{v}_2 = A\mathbf{u}_1 = \begin{pmatrix} -1/3 \\ -1/3 \end{pmatrix}$ $\mathbf{u}_2 = -3\mathbf{v}_2 = \begin{pmatrix} 1 \\ 1 \end{pmatrix}$

$\mathbf{v}_3 = A\mathbf{u}_2 = \begin{pmatrix} 3 \\ -2 \end{pmatrix}$ $\mathbf{u}_3 = \frac{1}{3}\mathbf{v}_3 = \begin{pmatrix} 1 \\ -2/3 \end{pmatrix}$

$\mathbf{v}_4 = A\mathbf{u}_3 = \begin{pmatrix} -1/3 \\ -1/3 \end{pmatrix}$ $\mathbf{u}_4 = -3\mathbf{v}_4 = \begin{pmatrix} 1 \\ 1 \end{pmatrix}$

(b) **Hint.** How are the eigenvalues of A related?

6. (a) **Hint.** If $\mathbf{x}_j$ is an eigenvector of A belonging to λ_j, then

$$B^{-1}\mathbf{x}_j = (A - \lambda I)\mathbf{x}_j = (\lambda_j - \lambda)\mathbf{x}_j = \frac{1}{\mu_j}\mathbf{x}_j$$

7. (b) **Hint.** It follows from part (a) that

$$(\lambda - a_{ii})x_i = \sum_{\substack{j=1 \\ j\neq i}}^{n} a_{ij}x_j$$

12. (a) **Solution.** Algorithm for computing eigenvectors of an $n \times n$ upper triangular matrix with no multiple eigenvalues.

      Set $U_1 = (1)$
        For $k = 1, \ldots, n-1$
          Use back substitution to solve

$$(R_k - \beta_k I)\,\mathbf{x}_k = -\mathbf{b}_k$$

where

$$\beta_k = r_{k+1,k+1} \quad \text{and} \quad \mathbf{b}_k = (r_{1,k+1}, r_{2,k+1}, \ldots, r_{k,k+1})^T$$

Set

$$U_{k+1} = \begin{bmatrix} U_k & \mathbf{x}_k \\ \mathbf{0}^T & 1 \end{bmatrix}$$

End (For Loop)

The matrix U_n is upper triangular with 1's on the diagonal. Its column vectors are the eigenvectors of R.

(b) **Solution.** All of the arithmetic is done in solving the $n-1$ systems

$$(R_k - \beta_k I)\mathbf{x}_k = -\mathbf{b}_k \qquad k = 1, \ldots, n-1$$

by back substitution. Solving the kth system requires

$$1 + 2 + \cdots + k = \frac{k(k+1)}{2} \quad \text{multiplications}$$

and k divisions. Thus the kth step of the loop requires $\frac{1}{2}k^2 + \frac{3}{2}k$ multiplications/divisions. The total algorithm requires

$$\frac{1}{2}\sum_{k=1}^{n-1}(k^2 + 3k) = \frac{1}{2}\left[\frac{n(2n-1)(n-1)}{6} + \frac{3n(n-1)}{2}\right]$$

$$= \frac{n^3}{6} + \frac{4n^2 - n - 4}{6} \quad \text{multiplications/divisions}$$

The dominant term is $n^3/6$.

7 LEASTSQUARES PROBLEMS

Overview

In this section we study computational methods for finding least squares solutions to overdetermined systems. In particular we discuss and compare methods based on the normal equations, the Householder QR factorization, and the singular value decomposition. The latter approach is the method of choice for rank deficient least squares problems. In this case there are infinitely many least squares solutions. Using the singular value decomposition we can find a *pseudoinverse* of the coefficient matrix and this can be used to compute the particular solution that has the smallest 2-norm.

Important Concepts

1. **Pseudoinverse of a matrix.** If A is an $m \times n$ matrix of rank r with singular value decomposition $U\Sigma V^T$, then the matrix Σ will be an $m \times n$

matrix of the form

$$\Sigma = \left(\begin{array}{c|c} \Sigma_1 & O \\ \hline O & O \end{array} \right) = \left(\begin{array}{cccc|c} \sigma_1 & & & & \\ & \sigma_2 & & & O \\ & & \ddots & & \\ & & & \sigma_r & \\ \hline & O & & & O \end{array} \right)$$

The *pseudoinverse* of A, denoted A^+, is defined by

$$A^+ = V\Sigma^+ U^T$$

where Σ^+ is the $n \times m$ matrix

$$\Sigma^+ = \left(\begin{array}{c|c} \Sigma_1^{-1} & O \\ \hline O & O \end{array} \right) = \left(\begin{array}{cccc|c} \frac{1}{\sigma_1} & & & & \\ & \ddots & & & O \\ & & \frac{1}{\sigma_r} & & \\ \hline & O & & & O \end{array} \right)$$

Important Theorems and Results

▶ **The Householder** QR **method.** If A is an $m \times n$ matrix with full rank, the least squares problem can be solved as follows:

1. Use Householder transformations to compute
$$R = H_n \cdots H_2 H_1 A \qquad \text{and} \qquad \mathbf{c} = H_n \cdots H_2 H_1 \mathbf{b}$$
where R is an $m \times n$ upper triangular matrix.

2. Partition R and $\mathbf{c}$ into block form:
$$R = \left(\begin{array}{c} R_1 \\ O \end{array} \right) \qquad \mathbf{c} = \left(\begin{array}{c} \mathbf{c}_1 \\ \mathbf{c}_2 \end{array} \right)$$
where R_1 and $\mathbf{c}_1$ each have n rows.

3. Use back substitution to solve $R_1 \mathbf{x} = \mathbf{c}_1$.

▶**The Penrose conditions.** The pseudoinverse of a matrix A is the unique matrix X that satisfies the following conditions.

1. $AXA = A$
2. $XAX = X$

3. $(AX)^T = AX$

4. $(XA)^T = XA$

▶ THEOREM 7.7.1. *If A is an $m \times n$ matrix of rank $r < n$ with singular value decomposition $U\Sigma V^T$, then the vector*

$$\mathbf{x} = A^+\mathbf{b} = V\Sigma^+U^T\mathbf{b}$$

minimizes $\|\mathbf{b} - A\mathbf{x}\|_2^2$. *Moreover, if $\mathbf{z}$ is any other vector that minimizes* $\|\mathbf{b} - A\mathbf{x}\|_2^2$, *then* $\|\mathbf{z}\|_2 > \|\mathbf{x}\|_2$.

Exercises: Solutions and Hints

3. (a) **Solution.**

$$\alpha_1 = \|\mathbf{a}_1\| = 2, \quad \beta_1 = \alpha_1(\alpha_1 - a_{11}) = 2, \quad \mathbf{v}_1 = (-1,\ 1,\ 1,\ 1)^T$$

$$H_1 = I - \frac{1}{\beta_1}\mathbf{v}_1\mathbf{v}_1^T$$

$$H_1 A = \begin{pmatrix} 2 & 3 \\ 0 & 2 \\ 0 & 1 \\ 0 & -2 \end{pmatrix} \qquad H_1\mathbf{b} = \begin{pmatrix} 8 \\ -1 \\ -8 \\ -5 \end{pmatrix}$$

$$\alpha_2 = \|(2,\ 1,\ -2)^T\| = 3 \qquad \beta_2 = 3(3-2) = 3 \qquad \mathbf{v}_2 = (-1,\ 1,\ -2)^T$$

$$H_2 = \begin{pmatrix} 1 & \mathbf{0}^T \\ \mathbf{0} & H_{22} \end{pmatrix} \text{ where } H_{22} = I - \frac{1}{\beta_2}\mathbf{v}_2\mathbf{v}_2^T$$

$$H_2 H_1 A = \begin{pmatrix} 2 & 3 \\ 0 & 3 \\ 0 & 0 \\ 0 & 0 \end{pmatrix} \qquad H_2 H_1\mathbf{b} = \begin{pmatrix} 8 \\ 0 \\ -9 \\ -3 \end{pmatrix}$$

5. Hint. Show first that if $A = \Sigma$, then Σ^+ satisfies the four Penrose conditions.

7. Hint. If $X = \dfrac{1}{\|\mathbf{x}\|_2^2}\mathbf{x}^T$, then

$$X\mathbf{x} = \frac{1}{\|\mathbf{x}\|_2^2}\mathbf{x}^T\mathbf{x} = 1$$

Using this it is easy to verify that $\mathbf{x}$ and X satisfy the four Penrose conditions.

9. Solution. If

$$\mathbf{b} = AA^+\mathbf{b} = A(A^+\mathbf{b})$$

then it follows that $\mathbf{b} \in R(A)$ since

$$R(A) = \{A\mathbf{x} \mid \mathbf{x} \in R^n\}$$

Conversely if $\mathbf{b} \in R(A)$, then $\mathbf{b} = A\mathbf{x}$ for some $\mathbf{x} \in R^n$. It follows that

$$A^+\mathbf{b} = A^+A\mathbf{x}$$
$$AA^+\mathbf{b} = AA^+A\mathbf{x} = A\mathbf{x} = \mathbf{b}$$

10. Solution. A vector $\mathbf{x} \in R^n$ minimizes $\|\mathbf{b} - A\mathbf{x}\|_2$ if and only if $\mathbf{x}$ is a solution to the normal equations. It follows from Theorem 7.7.1 that $A^+\mathbf{b}$ is a particular solution. Since $A^+\mathbf{b}$ is a particular solution it follows that a vector $\mathbf{x}$ will be a solution if and only if

$$\mathbf{x} = A^+\mathbf{b} + \mathbf{z}$$

where $\mathbf{z} \in N(A^TA)$. However, $N(A^TA) = N(A)$. Since $\mathbf{v}_{r+1}, \ldots, \mathbf{v}_n$ form a basis for $N(A)$ it follows that $\mathbf{x}$ is a solution if and only if

$$\mathbf{x} = A^+\mathbf{b} + c_{r+1}\mathbf{v}_{r+1} + \cdots + c_n\mathbf{v}_n$$

13. Hint. Show first that the three conditions hold for the case $A = \Sigma$.

MATLAB EXERCISES

3. (a) Since L is lower triangular with 1's on the diagonal, its determinant is 1. It follows that

$$\det(C) = \det(L)\det(L^T) = 1$$

and hence $C^{-1} = \operatorname{adj}(C)$ (See Section 3 of Chapter 2). Since C is an integer matrix its adjoint will also consist entirely of integers.

7. Hint. Look at the row sums of $A - tI$. Why must the row vectors of $A - tI$ be linearly dependent?

9. (b) $\operatorname{Cond}(X)$ should be on the order of 10^8, so the eigenvalue problem should be moderately ill-conditioned.

10. (b) $K\mathbf{e} = -H\mathbf{e}$.

12. (a) The graph has been rotated $45°$ in the counterclockwise direction.

(c) The graph should be the same as the graph from part (b). Reflecting about a line through the origin at an angle of $\frac{\pi}{8}$ is geometrically the same as reflecting about the x-axis and then rotating 45 degrees. The later pair of operations can be represented by the matrix product

$$\begin{pmatrix} c & -s \\ s & c \end{pmatrix} \begin{pmatrix} 1 & 0 \\ 0 & -1 \end{pmatrix} = \begin{pmatrix} c & s \\ s & -c \end{pmatrix}$$

where $c = \cos\frac{\pi}{4}$ and $s = \sin\frac{\pi}{4}$.

13. (b)

$$\mathbf{b}(1,:) = \mathbf{b}(2,:) = \mathbf{b}(3,:) = \mathbf{b}(4,:) = \tfrac{1}{2}(\mathbf{a}(2,:) + \mathbf{a}(3,:))$$

(c) Both A and B have the same largest singular value $s(1)$. Therefore

$$\|A\|_2 = s(1) = \|B\|_2$$

The matrix B is rank 1. Therefore $s(2) = s(3) = s(4) = 0$ and hence

$$\|B\|_F = \|\mathbf{s}\|_2 = s(1)$$

14. (b)

$$\|A\|_2 = s(1) = \|B\|_2$$

(c) To construct C, set

$$D(4,4) = 0 \quad \text{and} \quad C = U * D * V'$$

It follows that

$$\|C\|_2 = s(1) = \|A\|_2$$

and

$$\|C\|_F = \sqrt{s(1)^2 + s(2)^2 + s(3)^2} < \|\mathbf{s}\|_2 = \|A\|_F$$

16. (b) The disk centered at 50 is disjoint from the other two disks, so it contains exactly one eigenvalue. The eigenvalue is real so it must lie in the interval $[46, 54]$. The matrix C is similar to B and hence must have the same eigenvalues. The disks of C centered at 3 and 7 are disjoint from the other disks. Therefore each of the two disks contains an eigenvalue. These eigenvalues are real and consequently must lie in the intervals $[2.7, 3.3]$ and $[6.7, 7.3]$. The matrix C^T has the same eigenvalues as C and B. Using the Gerschgorin disk corresponding to the third row of C^T we see that the dominant eigenvalue must lie in the interval $[49.6, 50.4]$. Thus without computing the eigenvalues of B we are able to obtain nice approximations to their actual locations.

CHAPTER TEST A

1. **Hint.** Make up an examples using 2-digit decimal floating point arithmetic.
2. **Hint.** Make up some examples where not all of the matrices are square matrices.
3. **Hint.** Look at some of the examples in Section 4.
4. **Hint.** See the remarks following Theorem 7.6.1. in the text.
5. **Hint.** If the matrix is nonsymmetric then the eigenvalue problem could be ill-conditioned.
6. **Hint.** See Exercise 10 of Section 2.
7. **Hint.** See Theorem 7.4.2.
8. **Hint.** Look at some diagonal matrices.
9. **Hint.** See the discussion of the normal equations in Section 7.
10. **Hint.** Let A an $m \times n$ whose smallest singular value satisfies $0 < \sigma_n < \epsilon$ and let B be the best rank $n - 1$ approximation to A.

CHAPTER TEST B

1. **Hint.** The computation of $(AB)\mathbf{x}$ involves a matrix-matrix multiplication followed by a matrix-vector multiplication. The multiplication $A(B\mathbf{x})$ should be more efficient since it only involves 2 matrix-vector multiplications.

2. (a) **Solution.**

$$
\begin{pmatrix}
2 & 3 & 6 & | & 3 \\
4 & 4 & 8 & | & 0 \\
1 & 3 & 4 & | & 4
\end{pmatrix}
\rightarrow
\begin{pmatrix}
4 & 4 & 8 & | & 0 \\
2 & 3 & 6 & | & 3 \\
1 & 3 & 4 & | & 4
\end{pmatrix}
\rightarrow
\begin{pmatrix}
4 & 4 & 8 & | & 0 \\
0 & 1 & 2 & | & 3 \\
0 & 2 & 2 & | & 4
\end{pmatrix}
\rightarrow
$$

$$
\begin{pmatrix}
4 & 4 & 8 & | & 0 \\
0 & 2 & 2 & | & 4 \\
0 & 1 & 2 & | & 3
\end{pmatrix}
\rightarrow
\begin{pmatrix}
4 & 4 & 8 & | & 0 \\
0 & 2 & 2 & | & 4 \\
0 & 0 & 1 & | & 1
\end{pmatrix}
$$

Solving by back substitution we see that $x_1 = -3$ and $x_2 = x_3 = 1$.

(b) **Solution.**

$$
P = \begin{pmatrix}
0 & 1 & 0 \\
0 & 0 & 1 \\
1 & 0 & 0
\end{pmatrix}
\quad
L = \begin{pmatrix}
1 & 0 & 0 \\
\frac{1}{4} & 1 & 0 \\
\frac{1}{2} & \frac{1}{2} & 1
\end{pmatrix}
\quad
U = \begin{pmatrix}
4 & 4 & 8 \\
0 & 2 & 2 \\
0 & 0 & 1
\end{pmatrix}
$$

3. **Hint.** Show that the singular values of Q are all equal to 1.

5. **Hint.** If 15 digit decimal arithmetic is used the machine epsilon will be 0.5×10^{-14}.

7. **Hint.** $B = Q^T A Q$.